PIG HOUSING

PIG
HOUSING

DAVID SAINSBURY
M.A., B.Sc., Ph.D., M.R.C.V.S.
School of Veterinary Medicine, University of Cambridge

FARMING PRESS LTD
FENTON HOUSE, WHARFEDALE ROAD, IPSWICH

First published 1963
Second Impression 1964
Second Edition 1970
Third Edition 1972

SBN   85236   001   0

This book is set in 11 pt. on 13 pt. Times and printed in Great Britain on S.E.B. Antique Wove paper by Page Bros (Norwich) Ltd., Norwich

ACKNOWLEDGMENTS

I AM extremely grateful to the North of Scotland College of Agriculture for permission to reproduce their plans of the Craibstone ark, the adapted building for floor feeding with central-slatted dunging passage, the Muir of Pert rearing-house, sow-stalls, a slatted-floor feeding house, indoor multi-suckling pens and the pen gate. I also thank Mr J. W. Benham, F.R.I.C.S., 43, Duke Street Chelmsford, for reproduction of drawings of totally-enclosed fattening houses, the yarded piggery for fatteners, and sow yard with individual feeders. The fumigation procedure was devised by Dr E. G. Harry, of Houghton Research Station.

Many of the photographs were supplied by *Pig Farming*, but I also gratefully acknowledge those provided by: Andover Timber Co Ltd, farrowing crate; Black, Sivalls & Bryson (G.B.) Ltd, Bacon Bin fattening houses; BRC Engineering Co, Weldmesh dunging passage; Flower's Equipment Co, Nurs-Maid pig rearer; Vic Hallam Ltd, totally-enclosed fattening house; Harper Adams Agricultural College, Harper Adams pig parlour; Maywick Appliances Ltd, gas heater for creep; Pigloo Products, pig hut; Mr Anthony Phelps, of Liss, Hants, general pig housing pictures; Mr David Soutar, sow-stall house; Spillers Ltd, pig house on their farm at Middle Aston; Sir Richard Prince-Smith, MA, convertible farrowing and rearing pens; Mr Brian S. Thomas, modern intensive pig unit; Stock buildings Ltd, slatted-floor fattening pens.

ACKNOWLEDGMENTS

I should also like to pay tribute to the help I received from Mr L. A. Bunker, Mr A. S. Cray, Mr H. Hollinrake, Mr Seaton Baxter, Mr W. Marshall, Messrs G. & B. McGuckian, Mr M. J. Parker, Mr Richard Roadnight, Mr Norman Snell, Mr Philip Solari, Mr K. Thornton and Mr James Watson. To this group of people we shall always be grateful for developments of the first importance. In the same way we are deeply indebted to Mr W. T. Price and Mr John Luscombe for their services to the industry; reproduction has been made of certain of their designs including the Crark and Natural farrowing unit, not to mention fattening houses of Harper Adams inspiration.

I am also glad to acknowledge the important work of the Rev Peter Buckler and Silcock & Lever Feeds Ltd on the development of slatted floors, and on pig housing generally the invaluable efforts of Mr Clement Pointer, lately pig environment specialist in the Agricultural Development and Advisory Service and Mr Charles Dobson of the same Service, and the equally useful assistance rendered by the pig housing specialists of the Meat and Livestock Commission.

The plans on pages 96, 104, 128, 159, 161 and 163 are reproduced from "Housing the Pig" by kind permission of HM Stationery Office. The plans of the slatted floor farrowing house (page 86) and Mr. J. M. Anderson's feeding piggery with pressurised ventilation (page 162) are acknowledged to the Scottish Farm Buildings Investigation Unit.

Finally, I should like to express my personal thanks to Miss Margaret Pendle for technical assistance and to Mrs Anne Langley for help in the production of this book.

D.S.

CONTENTS

7

CONTENTS

APPENDICES

ILLUSTRATIONS

PLATES

DIAGRAMS

11

"BLUEPRINT" PIGHOUSE PLANS AND FITTINGS

FOREWORD

by Sir Richard Verdin, O.B.E., T.D., D.L., J.P.
Deputy Chairman of the Meat and Livestock Commission

THERE must be in practice nearly as many different methods of rearing pigs as there are pig farmers. Everyone has his own theories. Perhaps this is why there has been a growing need in this country for an independent and unbiased appraisal of pig housing and systems of management. This book, probably the most comprehensive ever written, satisfies that need. It is packed with sound common sense and valuable information about building materials and design, insulation and ventilation, equipment and fittings, labour-saving devices and disease control. It steers a wise course between theory and practice and reaches many conclusions which are of real value and use to the pig farmer.

Personally, I am particularly impressed by Dr Sainsbury's approach to farrowing houses and equipment. This aspect of management has become important through the industry's increasing awareness of the staggering losses which occur in young pigs. There is no doubt that a great many of these losses are due to bad management and bad housing. Dr Sainsbury is to be congratulated on pinpointing the best ways of avoiding cold, damp, draughty houses and inadequate facilities at farrowing time. His advice should help pig farmers to save money and pigs, not only at farrowing time, but throughout their lives.

This book will appeal to technical and practical people alike. It will be a text-book to student and teacher and a guide to amateur and professional. It is thoroughly readable and a mine of information on all aspects of this extremely important subject. I am sure you will find it fascinating and instructive.

Richard Verdin

Ridgmont Street,
London, WC1

13

DEDICATION

To DAVID SOUTAR, Wise Counsellor and Friend. A True Pioneer in Pig Housing and Inspirer of the Farm Buildings Association. Now leading the Scottish Farm Building Investigation Centre to a premier international position.

AUTHOR'S PREFACE

THE aim of this book is to present an up-to-date account of the principles and practice of pig housing. The topic is a particularly dynamic and progressive one and new developments are constantly taking place. There is a danger in writing a book on a subject which is changing so rapidly that the material may be out of date almost before it appears. I hope this has been avoided because the publishers have produced the book with great speed and efficiency and little time has elapsed between the writing and appearance of the book. I have tried to steer a careful course between the tried and tested on the one hand, and the most modern but relatively untried developments on the other. Where a design or idea is as yet too new to have been adequately used under varying practical conditions I have said so.

The history of specialised pig housing is remarkably short and one cannot expect in this time standardisation in methods, equipment or building techniques. Nevertheless, there is a great need for rationalisation to reduce the cost to the farmer and to give him a product that he is sure will function adequately. I believe, however, we are on the threshold of achieving this. Manufacturers are taking an increasing interest in this field. Advisers in the Agricultural Development and Advisory Service, Meat & Livestock Commission and in private practice are fulfilling an increasingly useful function, whilst there is the increasing pioneer work of the Farm Buildings Investigation

Unit of the Scottish Department of Agriculture under Mr David Soutar, continuing to place the entire pig industry in its debt. In addition, we have the Farm Buildings Department of the Agricultural Research Council's Institute at Silsoe actively pursuing research in pig housing and the Farm Buildings Centre and Pig Demonstration Area at the National Agricultural Centre, acting in an important manner as centres for investigation and the collection and dissemination of information. Dr L. E. Mount has built up a fine unit at the ARC'S Institute of Animal Physiology which is giving us a great deal of information on the fundamental physiological needs of the pig. If proper use is made of all the information now coming forward, then the best designs and features will be more definitely established and simplification of housing and other techniques can be expected. The saving to the farmer and to the industry will be considerable. I hope this book will provide a further impetus to this end.

In compiling this volume I have endeavoured to record the best of current practice in this country and abroad. This has left me greatly indebted to many farmers, scientists and advisers and in the text, or in the acknowledgment section, I have attempted to give credit in the many cases where it is due. I must, however, in addition thank the publishers, and in particular Mr Philip Wood and Mr Bernard Hogley, for their most untiring and efficient help, co-operation and encouragement. It has made my part a most enjoyable experience.

In my several years of involvement in this field I never cease to be impressed by the ingenuity and experimentation shown by the pig farmer. It is impossible to name all who have contributed their quota of ideas, but to them and to all others who are interested in this subject, I hope this book will at least form a basis for evolving new schemes, improvements and developments. To the reader who is less experienced, my hope is that the book will bring a little order into a subject that may often seem extremely confusing.

DAVID SAINSBURY

School of Veterinary Medicine,
University of Cambridge

16

AUTHOR'S PREFACE

THOUGH it is only two years since the last edition, advances in pig housing have moved on at their customary fast pace and it has been found necessary to carry out some important revisions on most of the practical aspects of pig housing. The most striking trends have been concerned with the development of larger units, further specialisation in design, automatic feeding arrangements, muck disposal and early weaning techniques.

I still live in the hope that it may one day be possible to report more *standardisation* as well as *specialisation* but I fear that, if anything, there has been a proliferation of new designs whilst most of the existing ones still have their usefulness.

After considerable thought I have retained the British rather than the metric system as the standard for measurements, though there is an easy-to-use conversion table available at the end of the book. It is inevitable that a change *will* have to take place, and if the past is anything to go by, a further revision will be required within two or three years, and this may be the most appropriate time for it.

If in reading this further edition I can succeed in infecting you with some of the enthusiastic and inventive pioneering of the leaders in pig housing development, then I shall have achieved much of the aim.

DAVID SAINSBURY

School of Veterinary Medicine,
Cambridge

B

Chapter 1

INTRODUCTION

THERE cannot be many aspects of animal husbandry that have aroused more interest, development and controversy in the last few years than the housing and environment required by pigs. A great deal of this arises from the appallingly bad state of affairs that existed in this field in the 1930's and 1940's.

It was at the beginning of this period that the pig industry first turned real attention to housing pigs intensively in large numbers. What catastrophies resulted! As I write, I can see a few fields away the roof of a Danish-type fattening piggery erected in 1935, which at the time was considered the last word in pig-housing design. Built to last a hundred years — and unfortunately it will do — it never reared pigs successfully, and after two years was abandoned as a piggery and has been used ever since as a store.

HEAVY LOSSES

This sort of thing was happening all over the country at the time. Reports were coming in of mortality rates in totally-enclosed piggeries of up to 40 to 50 per cent — and the more modern and splendid the design and the larger the size of the piggery, the more trouble it seemed to give. The lines of communication in agriculture were not as good then as they are now, and though pioneers, like the McGuckians in Northern Ireland, had personally solved their housing problems in pre-war days by means which we shall discuss later, the wrong sort of building on 1930 lines was still being commonly erected into the 1950's.

We cannot say that it is not occurring even now — vast sums of money are still being wasted on careless designs, unwieldy enterprises and unsatisfactory materials. Pioneers make their mistakes and always will, but the man who follows tried and tested housing forms need not at this stage do so.

The Pig's Needs

The people who erected the ' palaces ' of the thirties learned one lesson which it is difficult to over-emphasise or reiterate too often. It is this: *in designing a piggery or a conversion the needs of the pig must come first.* After these have been adequately dealt with, consider very carefully the needs of the stockman. Neither of these points is difficult to satisfy, but only *after* they have been adequately considered should one look into the question of labour, economy and such items as easy or automatic methods of feeding, dung cleaning, weighing, and so on. It is because of this importance that the first chapter in this book will be concerned with the environmental needs of the pig, for it is round these needs that the piggery should be built and run.

A point which one is always at pains to emphasise is that in designing or building a pig unit a high degree of pessimism is required! By this one implies that a farmer must ask himself how the housing can be planned to help him when and if there are outbreaks of disease. Experience shows that some of the most troublesome units are those where a degree of optimism little short of gambling has been resorted to. Large buildings on a restricted area have grown up and an outbreak of disease has led virtually to the collapse of the whole enterprise because disease in one pig passes with great ease to all the other pigs in the establishment, and unless the whole site is emptied the chain of infection cannot be broken. De-population involves emptying the whole unit to get rid of the trouble rather than just one or two buildings.

It will be noticeable in reading this book that the most successful designs strictly limit the number of pigs kept in close contact, and we find in practice that some of the most progressive farmers are following this trend. If you build on these lines, drug and veterinary bills have every chance of being kept to a minimum. Such designs also need not preclude the use of buildings enabling the correct environmental temperatures and other conditions to be economically maintained.

The Pigman

After giving our first consideration to the pig, the second and almost as important requirement is to care for the needs of the pigman. Actually the first essential here, albeit often neglected, is to provide him with personal accommodation of the highest order. So that we do not forget this I have included the outlines of a good farm cottage (fig. 1). The cost of construction is approximately £5 per square foot.

After this, the most important consideration is to provide the pigman with facilities to do his job with a minimum of crude manual effort — removing where possible the humping of food and dung. A good pigman is a skilled stockman whose time is far too valuable, for example, to spend half his day shovelling muck. For this reason the trend is progressing in automating movement of food and dung. There is, however, room for some caution on these lines. It is possible to so automate these processes that a pigman can look after many more pigs than he can adequately inspect and care for. It is possible also, if one is not careful, that the design is so made — no doubt very cheaply — that the pig is enclosed in a kennel or box that not only makes it difficult to inspect the pig, but it also makes it well nigh impossible to know what conditions the pigs *are* being subjected to. It is therefore a very good thing if the stockman is able to walk through the accommodation himself. Not only is his own impression to be respected, but he will, by looking at the pigs and seeing their behaviour, be able to size up the position immediately. A pig is a sensitive animal to environment and will react very strongly against unfavourable conditions.

21

The Site

With the modern intensive 'controlled environment' building it is not right, as some believe, to regard as unimportant its site and locality. In the first place the general location of the enterprise and the climatic region in which it is situated are of tremendous importance. For example, a higher standard of insulation and ventilation control are required in the North and eastern parts of the country than in the South and western districts. Considerable disappointments have resulted from copying, for example, a piggery that works well on a sheltered site in Cornwall

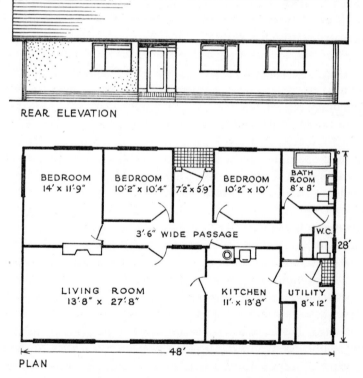

REAR ELEVATION

PLAN

Fig. 1. Plan of a modern three-bedroomed farm cottage.

22

and erecting it, say, on an exposed site in Lincolnshire. Piggeries must still be built very much with the locality in mind, in addition to the system of farming.

It is also desirable to erect a piggery on an open, well-drained site, with a southerly aspect. It is an advantage if the site is sloping slightly as this can help drainage and it may, in addition, assist with the construction of loading bays for food, fodder and livestock. Very clever use is made of siting in this way in Scandinavia and other more hilly countries than ours, and though it is not often possible to make full use of the advantages of steep hills in the British Isles it may often be more feasible than is sometimes apparent.

TAKE ADVANTAGE OF WINDBREAKS

Where buildings are on very exposed sites it is advisable to try and take advantage of wind-breaks from existing trees or to plant quick-growing trees if none is present; in any event, special care must be taken with ventilation both on exposed sites, where special baffling devices will be needed to reduce air flow, and on sheltered sites, where mechanical ventilation may be generally resorted to. Thus a much larger area of controlled inlet and cross ventilation area is needed (see chapter 4). It is also a good arrangement to have Dutch barns, implement sheds and stores on the north side of the pig-houses to form protection to the livestock buildings which will lie to the south.

The relationship of the various buildings must depend to a very great extent on the needs to reduce movement to a minimum and allow for mechanical handling of food, fodder and litter between and within the building. Many forms of mechanical handling are available, which are being more extensively utilised nowadays, together with bulk storage of food. Nevertheless, care must be taken that such necessities do not produce a unit where the stock buildings are so close that the disease risk is greatly increased and the ventilation is interfered with. For example, a good rule with piggeries is to allow a distance between the buildings themselves at least equal to the width of the buildings.

Where young stock are involved buildings should be periodically de-populated and where there is heavy density of stocking it is a wise procedure to allow a distance between sites of up to one mile. Even this will not preclude the passage of infections 'down wind'. It is, of course, very important to site buildings with good access to the road, but they must still be sited far enough back from the road — that is, 80 to 100 feet — to keep the building isolated from disease risk by stock passing along the road in trucks.

The Building

Pig buildings require a high degree of specialisation and few designs can be said to be adaptable for other purposes. By and large it is probably not unfair to say that the more adaptable is a pig-house for other purposes, the less efficient it is for pigs.

We have housing types to satisfy almost every demand of the numerous farming systems in the British Isles. It is perhaps the greatest strength of British pig-housing design that it is able to produce this multiplicity of systems, but it is also, at the same time, its weakness that the wrong house is so often used under the wrong circumstances.

In the specific sections for each type of pig-housing I shall try and make clear under what circumstances a particular form of housing is to be advocated. There are, however, a number of general requirements that are common to virtually any type of housing. Primarily, this springs from the fact that the environmental requirements of the pig are exacting and the housing must provide this at minimum capital and running costs.

HIGH EFFICIENCY ESSENTIAL

It is accepted that in the future we will probably be increasingly concerned with the production of pigs of high lean-meat content and high food-conversion efficiency. To achieve these most economically, high-efficiency housing is also essential. The cubic air space of the building or at least the lying area wants to be kept to a minimum so we can make the fullest use of the pig's body heat in warming the house. Likewise the insulation of the building must also almost invariably be good (there is one exception as will be seen on page 138) to prevent stresses from extremes and quick variation.

In order to provide the high standard of insulation that is required, methods have to be applied that are not known generally to the local builder; considerable knowledge therefore is required either by the farmer himself, or his professional adviser, on this factor (see page 46) or reliance may be placed on the manufacturer of prefabricated buildings. The last-named must, however, have considerable experience of pigs — and it *must be* pigs!

WITH OR WITHOUT STRAW

The usage of straw is a major item to be considered. In many areas, particularly in the western part of the British Isles, straw is expensive and difficult to obtain. It is therefore desirable to keep pigs without straw or other bedding except in the first few months of the pig's life, when shavings or sawdust may be the most satisfactory materials. When, however, no straw is used, the insulation of the floor and the surfaces generally must be impeccable. The dung and urine can be handled in liquid sludge form by automatic drainage combined with slatted floors. Though this system is most labour-saving, it involves relatively high capital outlay in arranging for the disposal of the liquid manure on the land; it is particularly suited to manuring grassland, but many pig farms have no outlet themselves and an arrangement may have to be made with neighbouring farmers.

If, on the other hand, straw is available on the farm and it is desired that the pigs should convert it into farmyard manure, another type of building is required and it usually means a form of house where the dung can be allowed to build up and be removed by mechanical means. If straw is to be used in large quantities, the pigs can make their own warm environment in the straw bed and the insulation of the structure of the house can be of a lower standard.

In planning the detail of the building, careful consideration must be given — and how seldom this does appear to be so — to the labour involved in the various chores that will go on in it. For example, in a farrowing house, consider feeding, cleaning, inspection, and veterinary attention to the sow; and feeding, cleaning, inspection, dosing, weighing, castration and veterinary attention to the piglets. In the fattening house look into the practices of feeding, cleaning, weighing, inspection and veterinary attention. It is certainly no trite statement to say that a few moments' thought on these matters in the planning stage may save hours every week of the piggery's existence.

A word may be appropriate at this early stage on the subject of the stocking density. Obviously, this has a very considerable bearing on the cost of pig housing. In pigs, just as with other livestock, the trend is towards high-density stocking in large and intensive units. But care must be taken not to overdo it. There is evidence that if pigs are stocked too densely, their productivity may fall quite markedly. Disease and vices can also be expected to rise in incidence the more pigs there are on a site unless the standard of management and hygiene increase in the same order.

FULL AUTOMATION NOT HERE YET!

I think some farmers consider that the most recent innovations in pig housing are beginning to take away the need for highly-skilled stock management, because feeding, cleaning out and ventilation can all be virtually automatic. Nothing could be further from the truth. All that the automatic processes can do is to relieve the stockman of wasteful manual effort which can be done better by machine. This leaves him more time for the actual care of the stock which no machine can ever replace.

Chapter 2

THE ENVIRONMENTAL NEEDS OF
THE PIG

PIGS have been kept intensively for many years and have shown themselves very responsive to the climatic susceptibility due to their relatively "naked" state, and this is especially noticeable in the early days of life when they are very vulnerable to cold, damp and draught. The young piglet has little to protect it from chilling, being poorly endowed with hair or subcutaneous fat and having a thin skin. Losses in pigs are high, generally running at 20–25 per cent[1, 2, 3, 4, 5], but surveys indicate that there is a marked seasonal effect, losses being at their worst from November to March. These losses [6, 7] can be alleviated by good housing (fig. 2). Moreover, the majority of losses are due to chilling and crushing; both factors that can be dealt with by good environmental control and housing. The vast majority of the losses due to poor environmental control occur in the first few days of life; thereafter the effects of poor environment are to cause not so much mortality and disease, as a loss in productivity, both in respect of liveweight gain and food conversion efficiency. By adulthood, pigs have become reasonably adaptable to a wide range of conditions which are not markedly different from those of other farm livestock.

Because of the difference between the various age groups, it is

27

appropriate to consider the climatic effects under three headings: piglets; growers and fatteners; and breeding pigs. In recent years there has been considerable controversy about the right conditions for pigs and a commentary on some of the experimental background to opinion is really necessary.

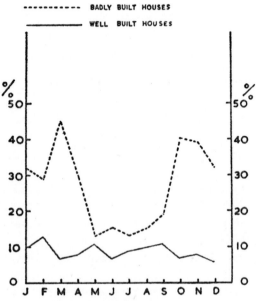

Fig 2. *Comparison of piglet losses throughout the year in badly-built and well-built houses.*

Piglets

Newly-born pigs have a poorly-developed heat-regulating ability, though their homothermic (i.e. temperature control) mechanisms develop quickly[8]. During the first hours after birth, the body temperature drop under cool conditions may be as much as 12°F, but the average will be approximately 4°F. The heavier piglet is able to withstand climatic change and cooling better than the small piglet. It is also much more able to resist crushing, and there is significant correlation between the weight of a pig and its ability to withstand cold stress.

Cairne and Pullar[9] measured the metabolic rate of pigs in a gradient-layer direct calorimeter at air temperatures of 59°F,

28

68°F, 77°F, and 86°F, the pigs in the experiment ranging from 9 to 26 lb. It was found that the 'critical' temperature of these pigs fed *ad lib* exceeded 86°F for pigs weighing 13 lb and under (say, up to two weeks), and was 68°F for pigs weighing 22 lb (say, four weeks).

EFFECT OF HEAT ON PIGLETS

It has been clear from the work done on young pigs that the environmental effects are not simple ones; that is, it is not, for example, just the simple question of temperature. Gill and Thompson [10] divided litters into equal groups and provided supplemental heat for half the litter to determine the effect of environment on suckling pigs. From weaning to 200 lb the pigs were fattened in the same environment. Their results showed that the piglets without additional heating were quite able to withstand cold though the average temperature was 44°F and even went below freezing on occasion. However, the pigs at the lower temperature consumed more solid food before weaning. Post-weaning performance was unaffected by the pre-weaning performance; the average time taken to reach 200 lb was 210 days for the heated group, and 213 for the unheated group; both, incidentally, poor results. It was still considered that there was sufficient saving in extra food consumed and in reduction of pigs crushed by the sow, to warrant the use of heat lamps through the first three weeks of the pig's life.

Pigs weaned at two weeks were reared under controlled conditions for three weeks in tests conducted by Muehling and Jensen [11]. Tests at temperatures of 40°, 50° and 60°F were used to compare the effects of heated houses, heat lamps and heat pads, and unheated houses. There were no significant differences in the rectal temperatures between treatments. Pigs in pens with an unheated draught-free house gained nearly as fast at 40°F as pigs in pens with heat. In a draughty building, however, where the temperature averaged 50°F but varied considerably, even pigs with access to the various heated areas did poorly.

This experiment emphasises two strikingly important requirements for the young pig—a complete absence of draughts and even, stable conditions that the piglet is adapted to. Few pig farmers lay enough emphasis on either of these.

ENVIRONMENT INFLUENCES GROWTH

McLagan and Thompson [12] farrowed and reared pigs in four different natural environments: (a) open pens in a large draughty granite-and-concrete building with an uninsulated floor without a damp-proof course, (b) in the same building as (a) but with a wooden sleeping platform, (c) a wooden ark hut with an indoor run, and (d) a wooden ark with an outdoor run. Pigs raised in environment (a) had average weaning weights *less than half as much* as those farrowed and raised in environment (d) even though the air-temperature difference between these two 'best' and 'worst' environments was only 2°F and could hardly account for the great difference (16·9 lb against 41·3 lb in one experiment).

The addition of the wooden sleeping platform in the open pen provided a dry bed for the pigs in that pen and their average weaning weight was more than twice that of the pigs in the open pen and nearly as much as the pigs in the best environment. Environmental factors other than air temperature clearly influence growth and appetite of the pigs. Though the open pen with the wood floor had the lowest temperature of all pens during the tests, the weaning weight of these pigs was nearly the heaviest and it seems that the floor type is of the greatest importance even if the house temperature drops to 45°F.

It is clear that we can therefore add dampness to draughts and changes in conditions as environments that are to be avoided for the young pig. The inter-relationships of temperature and air movement in their effects on piglet comfort have been shown by the author (fig. 3), indicating that with a high air velocity even a relatively warm temperature may be uncomfortable and conversely a lower air temperature but low air velocity can be comfortable.

Campbell and Paterson [13] studied the effects on the farm of giving supplementary heating to piglets prior to weaning. They found that the benefits were 2·41 lb extra weaning weight and

AIR VELOCITY

		Below 30ft/min	30–50ft/min	50–70ft/min
TEMPERATURE	70°F	Pigs of all ages comfortable	Pigs of all ages comfortable	Young piglets (1–8 weeks) uncomfortable
	65°F	Pigs below 1 week uncomfortable	Pigs below 5 weeks uncomfortable	Pigs below 12 weeks uncomfortable
	60°F	Pigs below 10 days uncomfortable	Young piglets (1–3 weeks old) uncomfortable	Piglets below 12 weeks uncomfortable
	55°F	Piglets below 8 weeks uncomfortable	Pigs below 12 weeks uncomfortable	—
	50°F	Pigs below 15 weeks uncomfortable	Pigs below about 16 weeks uncomfortable	—
	45°F	Pigs below 20 weeks uncomfortable	Pigs below 14 weeks uncomfortable	Pigs below 20 weeks uncomfortable
	40°F	Pigs below 20 weeks uncomfortable	Pigs below 20 weeks uncomfortable	Pigs below 20 weeks uncomfortable
	35°F	All fattening pigs uncomfortable		

Fig. 3. Temperature/air velocity chart showing "comfort" of pigs at various ages.

0·58 extra pigs weaned per litter compared with the unheated controls. They also pointed out that farmers often introduced bright emitter infra-red lamps into the unheated 'controls' or even built pens with warm-heated beds as being the only means of keeping the piglets alive; a bright light serves to attract the piglets initially to the warmed area but may not be necessary thereafter, as the piglets will return to the warmed nest whether or not the light bulb is switched on or off, once they have been shown the area.

HEAT OUTPUT OF PIGS

The heat output of pigs has been summarised by Bianca and Blaxter[14]. Their useful paper shows that (a) new-born pigs have low heat production rising more slowly than calves or lambs, and (b) thermal conductances are high at birth and fall with age.

Critical temperatures for piglets are given as 95°F at birth, 85°F up to 9 lb liveweight and 75°F at 22 lb liveweight at low air movements—less than 30 ft per minute. However, it should be noted that these temperatures are for single pigs, and in practice piglets are grouped together, hence the lower temperatures that seem all right in practice. Mount[15] has shown that the changes in metabolic rate of a *group* of young pigs exposed to falling environmental temperature tend to correspond with those of a *single* pig of the same total weight. Similarly, Mount[16] and Mount, Fuller, Hosie and Ingram[17] gave further details of the piglets' response to environment with many practical implications. For example, heat loss from groups of pigs exhibits a 24-hour cycle, with a maximum rate in the late afternoon and a minimum rate in the early morning, with a mean amplitude of the cycle at 68°F being 20 per cent of the mean value. Conductive heat loss of the piglet to the floor was measured by Mount[18] and also the partitioning of the heat loss of baby pigs between the different components.

Evaporative heat loss accounts for only about 10 per cent of the heat loss. The partitioning of the remaining 90 per cent is dependent on the floor. On concrete, 15 per cent was to the floor, 40 per cent by radiation, and 35 per cent by convection. On wood, 6 per cent was to the floor, 46 per cent by radiation and 30 per cent by convection. Substituting $\frac{1}{2}$ inch of wood for 1 inch of concrete was equivalent to raising the floor temperature 21·6°F. Raising the air speed from 20 to 60 feet per minute was equivalent to lowering the air temperature 9°F.

These findings agree with the observations made by the author regarding the comfort level of piglets and given in fig. 3 which show that by doubling air velocities the air temperature requirements rise 10°F. All these facts emphasise the need for straw, shavings or other bedding to be placed over the concrete of the floor, and the necessity for warmth and freedom from high air movement. With these exacting provisions successfully made the piglets are encouraged to lie away from the sow and the danger of crushing is greatly reduced.

Growers and Fatteners

The most comprehensive work on fattening pigs has been conducted in controlled-environment conditions by Californian workers[19, 20]. They studied the effect of temperatures between 40° and 110°F with a relative humidity of about 50 per cent and constant air velocity of 25 to 35 feet per minute. The daily gain reached a maximum at 73°F for 100-lb pigs, reducing to 65°F approximately at 200–210 lb (bacon weight). The rate of gain was severely reduced when air temperature varied from the optimum with the most rapid reduction occurring with the air temperature above the optimum. When air temperature was 100°F all pigs at 150 lb or over lost weight. The daily rate of gain of any size of pig tested was 2.0 lb or more at the optimum temperature. It was also found that at the temperature at which the maximum weight gain took place, the food was converted at its maximum efficiency (fig. 4).

Though this data is probably the most reliable, there are a number of conflicting reports that suggest the range in practice may be wider than indicated by the Californian workers. A summary of the general position can therefore be fairly stated in the following way. Animals penned in groups, as opposed to being penned separately, can thrive at lower house temperatures due to the effects of huddling and the reduction of radiation loss. It is clear that below 40°F increased food consumption and increase in physical exertion takes place, but above 40°F and below 85°F pigs may apparently thrive.

Data from Scandinavia[21] show that with adequate straw bedding pigs may do as well between 40° and 50°F as between 60° and 70°F, whilst from Northern Ireland[22] we find a temperature range of 80–85°F favoured, though most data suggest that above 75°F feed consumption begins to decline.

Whether the temperature for the fattener is constant or fluctuating ('cycling') seems to make little difference, provided that the mean of a cycling temperature is approximately that of

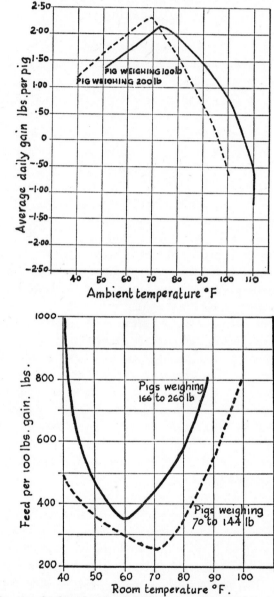

Fig. 4. *These graphs show the relationship between ambient (air) temperatures and the liveweight gain and feed conversion efficiency of pigs, referred to on page 33.*

the constant temperature and variation above or below this mean does not exceed approximately 10°F (20°F total) in a 24-hour period[23].

Moustgaard, Brauner, Neilson and Sørensen[24] suggested on the basis of their work that the optimum temperature of fatteners lies between 52°F and 68°F which is mid-way between the recommendation of the Californian and Ulster workers on the one hand and the Swedish evidence on the other. Essentially the work shows a considerable tolerance and acclimatisation of pigs to various temperatures, but it does not indicate that temperatures should be allowed to vary greatly diurnally. Food consumption also may not be too adversely affected provided the pig can have time to acclimatise to the temperature regime imposed upon it.

TYPES OF FLOOR

Inglis and Robertson[25] studied the actual heat loss through floors of various types, such as a pig might lie upon. They found that a wooden floor put over a concrete one cut the losses by half, whilst a thin layer of straw was only slightly less effective. The incorporation of hollow spaces and other insulating material in concrete floors had much smaller effects, but dampness had a very profound influence, wet concrete giving rise to 50 per cent more loss than dry. It would thus seem that where possible, plenty of bedding should be provided — particularly for the young pig — and always there should be a good damp-proof course and good drainage, to maintain as dry a surface as possible.

HEATED FLOORS

In recent trials at the Cambridge School of Veterinary Medicine it was found there was a significant improvement in the food conversion and growth rate of pigs kept on electrically-warmed floors over those in the control pens of ordinary insulated floors without bedding.

The results showed that the winter food conversion for pigs reared from 67–210 lb in the warmed pens was 3·45 lb feed per lb liveweight gain and in the control pen was 3·85 lb. The figures for growth rate in the winter were 1.04 lb per pig daily with the warm pen and 0·9 lb in the control pen.

This improvement in growth rate meant that the pigs in the warm pen would reach bacon weight (210 lb) 13 days sooner than the pigs in the control pen.

It is very clear from the amount of investigation on the needs of the fattener that temperature *per se* is not a good indicator of the comfort and optimum environmental state for keeping fattening pigs—any more than it is for piglets. Where a condition of warmth is needed consideration must be given equally to air temperature, air velocity (draught), humidity and radiation from surrounding surfaces.

'TURKISH BATH' PIGGERIES

As a contrast to the general belief in the importance of dryness with the warmth, Gordon and Luke[26, 27] have described graphically the 'Turkish Bath' atmosphere successfully followed in some of the Northern Ireland piggeries where conditions are kept at around 75° to 85°F and the relative humidity approaches 100 per cent. Only 4 to 5 square inches of ventilation space per pig are provided.

This moist climate has physiological advantages in the prevention and control of respiratory infections and the high ventilation and 'sedimentation' rates produced by the humidity lead to a good degree of air purity in the atmosphere. Gordon found that the bacteria-carrying particles in 'Turkish Bath' houses tend to be fewer in number and larger in size than in normal piggeries. Thus they are more effectively trapped in the nose, which in part explains the low incidence of coughing and pneumonia. The success of this system can therefore be attributed to its environment, which is highly antagonistic to disease-causing agents allied to the skill of the stockman in controlling these conditions so they do not become lethal to the pigs. This has been confirmed in some interesting comparative studies of housing systems by O'Grady *et al* in Ireland[28], where they found that high humidities reduced the incidence of pneumonia.

In addition, important work has been done on the interaction between environment and level of feeding[29] for pigs from weaning to bacon weight. It was found that the rates of growth and efficiencies of food conversion between 45 and 100 lb liveweight of pigs kept to a low plane of feeding were most adversely affected during the winter months by bad housing conditions wherein the average air temperature was 43°F. On the other hand, this bad housing had no adverse effect upon pigs of a similar size kept to a higher plane of feeding.

In the summer season, with its higher environmental temperatures in both good and bad piggeries, there was little difference between groups — but it is noteworthy that the trials were conducted in Aberdeenshire where high air temperatures in excess of the known upper limit a fattener can tolerate would not be found. It was also observed that during the growth period from 100 to 200 lb liveweight the efficiency of food conversion in the well-housed, low-plane diet pigs was significantly better than that for the well-housed, high-plane pigs, but was not significantly better than the efficiency of food conversion of the high-plane pigs from the bad piggery.

ENVIRONMENT AND CARCASE QUALITY

In the work of Moustgaard et al already referred to it was found that the highest protein deposition took place in an approximate temperature range of 59°-73°F which was the temperature giving fastest growth and lowest food conversion rates. Fuller[30], with single pigs in contrast to Moustgaard's group of pigs, found that nitrogen retention was highest at 77°F and lowest at 50°F and 86°F, which is in good agreement with Moustgaard's studies, bearing in mind that single pigs will lose heat more readily and thus require a higher temperature.

Some work has been done on the influence of the environment on carcase quality. Moustgaard et al showed that at low air temperatures the nitrogen retention was reduced relatively more than the rate of gain so that the carcase became fatter. It seems from this work that the least fatty carcase is produced at about 60°F with 70 per cent humidity. Both lower and higher temperatures, and higher humidities, tended to give fatter carcases in baconers. Holme and Coey[31] found that baconers penned individually 'did' better and had better carcase length at a temperature of 70°F over the weight range 40–200 lb than at 54°F whilst Smith and Tonks[32], with *ad lib* pigs between 50 and 200 lb, compared 82°F and 92 per cent relative humidity with 70°F and 70 per cent relative humidity. No effect resulted on the dressing-out percentage, carcase length, depth of backfat, colour and pH of eye-muscle, or iodine number of the fat, though eye-muscle area was slightly better at the lower temperature, and at the higher temperature and humidity there was a 27 per cent slower growth rate.

Lighting Needs of Pigs

Braude *et al* investigated the effect of light on fattening pigs. They tried the following treatments: continuous darkness, that is 24 hours total darkness; periods of 14 hours light and 10 hours darkness; 10 hours darkness and 10 hours light; and 24 hours light. In terms of weight gain or feed per lb gain, there were no significant differences in any treatment. There is thus much to be said for keeping fattening pigs in dim conditions, provided there is sufficient light for management, feeding, and for the maintenance of clean habits. The absence of windows prevents heat loss through bad insulation of glass and simplifies the construction of the house.

Similarly, negative findings on the effect of light on pigs have been recorded by other workers, though Scholz[34] did find a slightly favourable effect by darkening the pens, in so far as feed conversion was improved by 3 per cent and liveweight gain by 4 per cent.

Recently, however, Russian investigations[35] have shown the importance of the length of the lighting period for breeding females. Their work showed that gilts given an 18-hour light period per day exhibited a stronger, longer and more regular oestrus than gilts exposed to a 6-hour lighting period. They also compared the effects on gilts of 17-hour, continuous, and natural lighting conditions, the last being not dissimilar from the northern parts of Britain in winter. The Russian workers found that gilts exposed to the longer periods produced from 0 8 to 2·7 more piglets in the first litter than the controls, so that it can be recommended that 17-hour illumination is given for gilts for 10 to 20 days before mating and during gestation.

There is no doubt of the practical implications of this work because in recent times breeding sows have often been kept under dim conditions at most times. Whilst the Russians have not found definite evidence of an effect on sows, as opposed to gilts, there is a possibility that some of the breeding troubles in intensively-kept sows may be due to the poor attention given to lighting. A longer and brighter period of light should be given during suckling and for at least 25 days after service.

Summarising, it would seem that while the temperature requirements of the grower (or 'fattener') are well defined for the individually-housed pig without bedding, a wider tolerance is acceptable in practice where acclimatisation, grouping effects, and bedding can alter the picture considerably. Nevertheless, the limits are reasonably well defined for temperature, but this is not so in the case of air movement and humidity on which less work has been done. The effects of humidity appear to be associated more with an indirect one of disease, but a more accurate definition of these effects is needed. A great deal more study is also required on the use of supplementary heating where the picture is confusing both in terms of the type of heat application required and the prime *economic* temperatures to maintain.

Breeding Pigs

Most of the work that has been done in connection with sows has been concerned with high temperatures. However, sows exposed to temperatures as high as 99°F for periods of about 8 days could still generally produce normal litters[36]. It seemed, however, that in-pig sows showed more stress than empty ones due to the increased metabolic load. Heitman's work already referred to indicated that the best temperature for the sow was not the best for the piglet; the litter thriving best at 80°F and the sow at 60°F approximately. Where temperatures commonly stay above 85°F it is clear that there is justification for the provision of cooling devices such as sprinklers, sprays and wallows, in addition to shade. Several reports from America confirm this. For example, Whatley in Oklahoma[37], farrowed an average of 2·35 more live pigs per litter by sprinkling a group of pregnant sows, in comparison with an untreated group. The average maximum daily temperature was 96°F.

However, to the farmer in temperate climates, the problems of high summer temperatures are acute only under the most exceptional circumstances and much work remains to be done on the climatic needs of the breeding pig. For example, the special needs of the pregnant sow under the less-generous rationing systems now advised, or the relatively immobile conditions of the sow stall. It is reasonable to assume that such animals would require higher temperatures than those kept, for example, in deeply-strawed accommodation under traditional feeding programmes. There is also no clear definition of the first few days after farrowing when it is known that the sow's metabolic rate is reduced and thus its susceptibility to chilling may be increased. A thorough investigation into this problem is called for as it may not only elucidate the environmental needs of the sow, but may also help towards an understanding of some of the important metabolic disorders affecting the lactating sow.

STOCKING DENSITIES

One final word is appropriate on modern problems of stocking density and numbers per pen. No-one, indeed, disputes the importance of stocking a building to its maximum capacity; a full pen of pigs is usually the cleanest and most comfortable group. Nevertheless, there is some evidence to show that one should at least proceed with caution in packing the pigs in. In experiments in America [38] space allowance tests were made on pigs kept at 5 square feet, 10 square feet and 20 square feet per head. The results were as follows:

Square feet per pig	Gain (lb) during experiment	Average daily feed	Feed per unit gain
5	89·1	5·34	4·09
10	92·2	5·23	3·86
20	98·5	5·22	3·69

Above: Piggery in course of erection using pre-fabricated sections for roof and walls. *Left:* Prefabricated wall; note smooth asbestos lining that is cheap, hygienic and durable.

Above: One form of roof insulation—a sandwich infill under a corrugated-asbestos roof with an asbestos underlining sheet.
Below: Vapour-seal paper in position over a roof. Note the over-lapping provided at the joints.

Roof space in a well-insulated piggery showing the vapour-sealed insulation being placed in position and the ventilating trunk carrying the stale air to the outside.

Here the insulation lies directly under the roof, but note that there is no compression of it at any point

Damp-course of medium-gauge polythene, 6′ wide, laid on top of a concrete base.

Insulated panels in position awaiting rendering. The panels are used for floor, wall and roof insulation and are of wood-wool.

The rendering being placed on a house built of wood-wool slabs, forming a well-insulated and hygienic surface.

Investigations on the effect of numbers of pigs per pen were conducted by the same workers and the results were as follows:

Number of pigs per pen	Gain (lb)	Average daily gain	Feed per unit gain
3	94·3	5·66	4·15
6	93·7	5·13	3·79
12	93·1	5·00	3·71

It is apparent from this work that fatteners benefit quite markedly from plenty of space and small pen-size. That this statement seems justified is shown by the work of Gehlbach, Becher, Coss, Harmon and Jensen[39], who conducted experiments with 600 fatteners *ad lib* fed on different types of floor in groups of up to 16 pigs at different stocking densities. They found that group size affected level of performance and that the best group size was governed by the size and weight of the pig and by the environment. A lower rate of gain took place at the higher stocking densities, which it was believed was due to lower food consumption due to heat stress. The feed conversion was not affected by the stocking density in these experiments, though this is not a general finding.

Scholz studied fatteners in groups of 1, 10, 25, 90, 210 and 530 pigs. His findings were that with restricted feeding regimes 30 pigs are a maximum and for *ad lib* feeding not more than 90. He found the best results of all were when groups stayed in the same pen throughout the fattening period with the pen size restricted by sliding sides. It is interesting that Scholz recommends very high stocking rates (2 square feet per pig to 90 lb, 3 square feet from 90–130 lb, 4 square feet for 130 to 175 lb and 5 square feet up to 240 lb), but high air movement rates were recorded in the pens of 100 feet per minute, which is about twice the usual rate in the sleeping area.

The detrimental effect of heavy stocking seems sufficiently pronounced at least to be noted carefully by the designer and farmer. About 12–20 pigs per pen is probably our ideal and is unlikely to be an unwise choice; extremely dense stocking may, however, retard the growth of pigs unless the environmental conditions are carefully maintained at the optimum.

REFERENCES

1. LONGWELL, A. (1952). A Survey of Losses in Pigs on New Zealand Farms. 1949–1951. *N.Z. Jl. Sci. Technol.*, **34A**, 294–305.
2. HUTCHINSON, H. D. (1954). Causes of Baby Piglet Mortality. *J. Amer. Sci.*, **13**, 1,023.
3. REID, J. (1954). Rearing Pigs. *Vet. Rec.*, **66**, 862–865.
4. BRAUDE, R., CLARKE, D. M. and MITCHELL, K. G. (1954). Analysis of the Breeding Records of a Herd of Pigs. *J. Agric. Sci.*, **45**, 19–27.
5. FRASER, A. F. (1966). Studies of Piglet Husbandry in Jamaica. *Br. Vet. J.*, **122**, 322–332.
6. GRACEY, J. (1955). Survey of Pig Losses in Northern Ireland. *Vet. Rec.*, **67**, (50) 984–990.
7. SAINSBURY, D. W. B. (1955). Impact of Environment on the Health and Productivity of Pigs—Discussion. *Vet. Rec.*, **67**, (50), 1,011.
8. NEWLAND, H. W., MCMILLEN, W. N. and REINEKE, E. P. (1952). Temperature Adaptation in the Baby Pig. *J. Anim. Sci.*, **11**, 118.
9. CAIRNE, A. B. and PULLER, J. D. (1957). The Metabolism of the Young Pig. *J. Physiol.*, **139**, 15.
10. GILL, J. C. and THOMSON, W. (1956). Effect of Environmental Temperature on Suckling Pigs and a Study of the Milk Yield of the Sow. *J. Agric. Sci.*, **47**, 324.
11. MEULLING, A. J. and JENSEN, A. H. (1958). Paper given at Amer. Soc. of Agric. Engineers, December 1958, cited by T. E. BOND, *Agric. Engng.*, **40**, 548.
12. MCLAGAN, J. R. and THOMSON, W. (1950). Effective Temperature as a Measure of Environmental Conditions for Pigs. *J. Agric. Sci.*, **40**, 367.
13. CAMPBELL and PATERSON (1958). The Effect of Supplementary Heating on the Rearing of Pigs. *J. Agric. Sci.*, **51**, 234–236.
14. BIANCA, W. and BLAXTER, K. E. (1961). The Influence of the Environment on Animal Production and Health under Housing Conditions. Festschr. 8th Int. Tierzuchtkongr. Hamburg. 113–147.
15. MOUNT, L. E. (1960). The Influence of Huddling and Body Size on the Metabolic Rate of the Young Pig. *J. Agric. Sci.*, **55**, (1) 101–105.
16. MOUNT, L. E. (1966). Heat Loss from Young Pigs. Report. Inst. Anim. Physiol. Babraham. 1966. p. 46.
17. MOUNT, L. E., FULLER, M. F., HOSIE, K. F. and Ingram, D. L. (1961). Climatic Studies on Pigs. Report. Inst. Anim. Physiol. Babraham. 1960–1961. p. 23.
18. MOUNT, L. E. (1968) The Climatic Physiology of the Pig, 167–195. Edward Arnold, London
19. HEITMAN, H. and HUGHES, E. H. (1949). The Effects of Air Temperature and Relative Humidity on the Physiological Well-being of Swine. *J. Anim. Sci.*, **8**, 171–181.
20. HEITMAN, H., KELLY, C. F. and BOND, T. E. (1958). The Relation of Ambient Temperature to Weight Gain in Swine. *J. Amer. Sci.*, **17**, 62.
21. HELLBERG, A. (1961). The Reactions of Pigs to Low Temperature. Festschr. 8th Int. Tierzuchtkongr. Hamburg. 106–107.
22. GORDON, W. and LUKE, D. (1956). Observations on Restricted Ventilation in Pig Houses. *Vet. Rec.*, **68**, 1,030–1,031.
23. SØRENSON, P. H. (1961). The Influence of the Piggery Climate on the Growth, Food Utilisation and Carcase Quality of Pigs. Aarsberetn. Inst. Sterilitets forksn. Kgl. Vet. og. Landbokojsk. 185–201.

24. Moustgaard, J., Brauner, N., Sørenson, P. (1960). Influence of Environmental Temperature and Humidity on Growth, Feed Utilisation, and Bacon Quality of Pigs. Landlr. Bygg. Bygg-forskn-Inst. Kbh. **18**.
25. Inglis, J. S. S. and Robertson, A. (1953). The Measurement of Heat Loss through Floors. *Vet. Rec.*, **65**, 875–876.
26. Gordon, W. A. M. (1960). Environmental Studies in Pig Housing. Doctoral Thesis. Queen's University, Belfast.
27. Gordon, W. A. M. (1962). Environmental Studies in Pig Housing. *Brit. Vet. J.*, **118**, 171, 243.
28. O'Grady, J. F., Tuite, P. J., O'Brien, J. J. and Attwood, E. A. (1966).*Pig Farming*, **14**, 1.
29. Lucas, I. A. M. and Calder, A. F. (1955). The Interaction Between Environment and Level of Feeding for Pigs from Weaning to Bacon Weight.*J. Agric. Sci.*, **46**, 56.
30. Fuller, M. F. (1964). Ph.D. Thesis. University of Cambridge.
31. Holme, F. W. and Coey, W. E. (1966). Proceedings of the 44th meeting of the British Society of Animal Production.
32. Smith, W. C. and Tonks, H. M. (1966). Proceedings of 9th Internat. Congress of Anim. Prod.
33. Braude, R., Mitchell, K. G., Finn-Kelcey, P., Lowen, V. H. (1958). The Effect of Light on Fattening Pigs. *Proc. Nutri. Soc.*, **17**, 38.
34. Scholz, K. (1966). Proceeding of C.I.G.R. Second Section Seminar.
35. Pointer, C. G. (1972) Do Gilts and Sows need supplemetary lighting? *Pig International*, Jan 1972, 23–24
36. Kelly, C., Heitman, H. and Hughes, E. (1951). Effect of Elevated Ambient Temperature on Pregnant Sows. *J. Anim. Sci.*, **10**, 907–915.
37. Whatley, J., Palmer, J., Chambers, D., Stephens, D. The Value of Water Sprinklers for Cooling Pregnant Sows During Summer. *Misc. Publ. Okla. Agric. Exp. Sta.* No. 48, 1957. 2–4.
38. Bond, T. E., Heitman, H., Hahn, L., and Kelly, C. F. *California Agric.* (1962). 9–11.
39. Gehlbach, G. D., Becher, D. E., Coss, J. L., Harmon, B. G. and Jensen, A.H. (1966). *J. Anim. Sci.*, **25**, 386–391.

Chapter 3

PIGGERY CONSTRUCTION

THE principle feature required for all pig accommodation is to provide the correct environmental conditions as cheaply and economically as possible. This means, first and foremost, a high standard of thermal insulation of the surfaces all round the pig, including floor, walls, roof and windows. Nevertheless, it is as well in the first instance to see just where most of the heat is lost from a piggery so that the insulation can be applied in the most profitable manner.

Fig. 5 shows the loss of heat from each part of a typical uninsulated piggery, including that lost by ventilation, which is shown as a percentage of the total heat loss. The most interesting fact that emerges is that it is the roof through which most heat is lost, and indeed, this loss, together with that through ventilation, accounts for 80 per cent of the total. It should be noted that this does not include heat loss by conduction from the pig to any surface with which it has contact. This, of course, is difficult to assess but reference has been made to its importance in the previous chapter.

In fig. 6 we see what a vast improvement results if good principles of insulation are applied. The diagram shows the heat-loss through surfaces only in a moderately well insulated, and uninsulated piggery per 100-lb pig, contrasted with the heat production of a pig of this weight. The difference between inside and

44

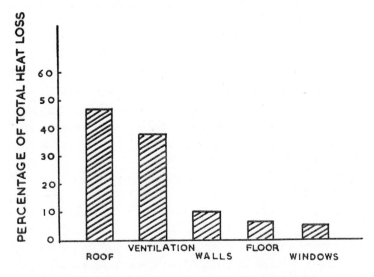

Fig. 5. *Heat losses from the parts of a typical uninsulated piggery.*

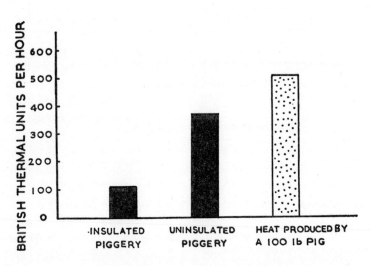

Fig. 6. *Heat losses through the surfaces of typical insulated and uninsulated piggeries with a difference of 10°F between inside and outside temperatures, contrasted with the heat production of a 100-lb pig.*

45

outside air temperatures have been taken as 10°F. To give the reader an idea of the quantities of heat involved it may be mentioned that seven pigs of this weight generate a similar amount of heat to that emitted by a one-bar electric fire. The heat production is therefore considerable but not enough to allow liberties to be taken either with the insulation or ventilation of the house.

It will also be seen from the diagram that, disregarding the loss of heat by ventilation, a difference of nearly 50°F between inside and outside temperatures is feasible in a well-insulated house and making use only of the pigs' own heat. Even allowing for the ventilation loss, a difference of 40°F is easily obtained, provided that the construction and ventilation are both on sound lines.

Good thermal insulation not only serves to retain the heat in winter, it also keeps the building cool in summer. It helps to prevent condensation and dampness, keeps any heating costs down and enables the farmer to maintain uniform and near-constant conditions in the house. The effects on stock are economically vital ones — by helping to maintain an optimum environment food costs are kept to a minimum and growth and good health are promoted.

Nevertheless, in spite of the importance of thermal insulation, it is a comparatively new idea in animal housing and has often been applied in a very unsatisfactory way. It is therefore worth looking into this matter in some detail so that the numerous mistakes of the past can be avoided.

Assessing Insulation Values

Before dealing with the strictly practical aspects of insulation we should have some knowledge of the way one can assess the respective insulation values of different materials or forms of construction. This will help us considerably in choosing our material.

First of all, attached to every building material, there is its thermal conductivity or "k" value. This figure is the amount of heat in British Thermal Units that is transmitted in one hour through one square foot of the material 1 inch thick, when the difference between the temperature of the inner and outer sides is 1°F. Therefore, the lower the "k" value the better the insulating qualities of a material.

In this way one can grade different materials according to their insulating qualities and it goes some way to answer the question as to which are the best insulators. A table of "k" values is given (fig. 7) and it can be seen, for instance, from this that you would need over twice the thickness of straw to give the same insulating value as a given thickness of glass or mineral wool though, in fact, both can be most useful insulators.

In fact "k" values are of limited use. For surfaces of pig-houses are generally composite structures. For example, an insulated roof might consist of an outer cladding of corrugated-asbestos sheets and an inner lining of mineral wool and fibreboard, and also an air space. What we really want to know is the rate of

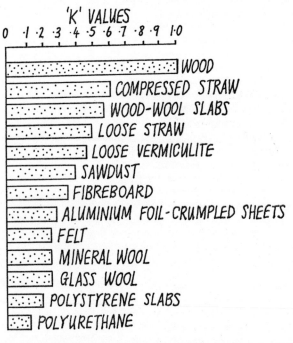

Fig. 7. " K" values of a series of materials used for thermal insulation. The size of blocks are proportional to heat losses through them so that the materials at the bottom are the best insulators.

heat loss (or heat gain during very hot summer weather) through the whole structure rather than just the individual materials.

This value takes us much further than the "k" value and is known as the "U" value. By definition, this is the amount of heat in British Thermal Units that is transmitted through one square foot of the construction from the air inside to the air outside when there is a 1°F difference in temperature between inside and outside. It is possible to build up the "U" value of a complete wall or roof structure if one has the "k" values of the individual materials (plus one or two other figures), but it is unnecessary to do this as "U" values for complete structures are available from the manufacturers.

As with "k" values, the lower the "U" value the better the insulating qualities. A table of representative values is given in fig. 8. The striking reduction in heat loss obtained by insulation is clearly seen, the black blocks being poor insulators and the cross-hatched blocks reasonable or good ones.

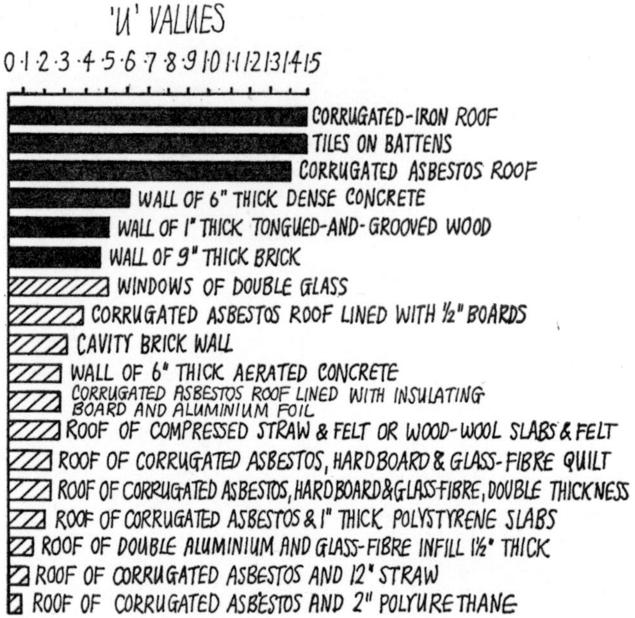

Fig. 8. " U " values of bad (solid blocks) and satisfactory or good (cross-hatched blocks) roof and wall constructions.

48

Before dealing with the practical aspects it is also worth setting ourselves a standard to build to. It is economic in most piggeries now to aim to have a "U" value in the roof of 0·1 or less. The cost of a high standard of insulation is very moderate. Mineral-wool insulation, for example, costs approximately 2p for each inch thickness, and even superb insulation of up to 3–4 inches thickness will not add more than about 6p per square foot to the cost.

For the walls, a figure between 0·2 and 0·3 is acceptable; with the floor, however, we should aim to be as near 0·1 as possible. It is important to eliminate condensation and the accompanying table shows whether the "U" value is able to do this for given differences of temperature between inside and outside.

'U' Value Required to Overcome Condensation

Maximum Temperature Difference (Internal-External)

		30	40	50	60	70
	60	0·80	0·59	0·47	0·40	0·33
	65	0·67	0·50	0·40	0·33	0·28
Relative	70	0·57	0·42	0·33	0·27	0·24
Humidity	75	0·47	0·34	0·27	0·22	0·19
%	80	0·35	0·27	0·21	0·18	0·15
	85	0·26	0·20	0·16	0·13	0·12
	90	0·19	0·14	0·13	0·10	0·08

In view of the amount of heat lost through the roof, this is obviously the appropriate place to start.

Insulating the Roof

An example of the correct construction of an insulated surface may, in fact, be taken by looking at the way one would insulate the roof — generally the weakest point in this respect in existing piggeries. The outer cladding, whether it is already present in an existing building or on a new house, is generally of corrugated asbestos, metal or tiles. All these are poor insulators — and quite unsatisfactory alone. We therefore place below these an insulation lining. This may follow the line of the roof if the pitch is shallow and the walls not too high. If, however, the roof is steep and the walls high, it is better by far to construct a flat "false" insulated ceiling at a height of around 6' 6". A very popular type

49

of insulating material to use is mineral or glass wool, and a correct thickness is 2"—though it is much preferable to use 3" or even 4". The "wool" insulation has to be suspended and a variety of materials can be used — for example, aluminium, fibreboard, plywood and flat asbestos sheets. Manufacturers use all these successfully, but the do-it-yourself builder will almost always use a water-repellant plywood, oil-tempered hardboard or asbestos sheet.

VAPOUR-SEAL

To stop moisture from within the house permeating the insulation it is essential to place a *vapour-seal* on the warm or underside of the insulation. Usually the sheets that suspend the insulation are permeable to water vapour, so the vapour-seal is placed immediately on top of the sheets. It may consist of polythene sheets or bituminised material such as sisalkraft paper. It is also

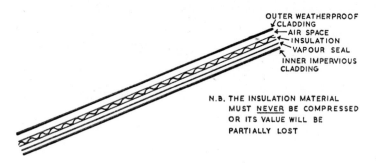

OUTER WEATHERPROOF CLADDING
AIR SPACE
INSULATION
VAPOUR SEAL
INNER IMPERVIOUS CLADDING

N.B. THE INSULATION MATERIAL MUST NEVER BE COMPRESSED OR ITS VALUE WILL BE PARTIALLY LOST

Fig. 9. Details of insulated roof construction.

possible to obtain mineral wool (Slagbestos) completely sealed in polythene bags and with a built-in overlapping piece. This is most useful for correct on-the-site work. The insulation material, whether mineral or glass wool, must never be compressed as its beneficial qualities depend upon the air spaces between the fibres remaining open.

From inside to outside the roof will therefore consist of the following layers: firstly the inner lining board, then the vapour-seal, followed by the insulating material. Alternative interior linings are as follows:

Fibreboards. As a roof lining an oil-tempered hardboard or an insulation fibreboard can be successful if applied in the manner recommended and if management ensures that the building is kept well ventilated. If these materials are likely to be abused in any way, they will not prove too satisfactory and

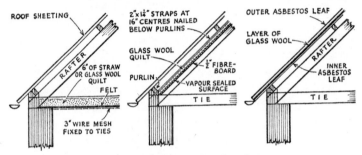

ROOF SHEETING

RAFTER

6" OF STRAW OR GLASS WOOL QUILT

FELT

3" WIRE MESH FIXED TO TIES

2"x1¼" STRAPS AT 16" CENTRES NAILED BELOW PURLINS

GLASS WOOL QUILT

PURLIN

½" FIBRE-BOARD

VAPOUR SEALED SURFACE

T I E

OUTER ASBESTOS LEAF

LAYER OF GLASS WOOL

RAFTER

INNER ASBESTOS LEAF

T I E

Fig. 10. Three forms of roof insulation.

are best avoided. Oil-tempered hardboard can also form an effective wall lining above pig height.

Plasterboards. An internal roof lining of plastic-faced plasterboards is used in several popular forms of house. This appears to be the only type of plasterboard that can be used, and all joints must be sealed completely with a plastic tape to ensure that no moisture reaches the board itself.

Note that both the above lining boards only give a fairly small measure of insulation and must be topped with a layer of glass-fibre or mineral-wool.

Asbestos-based Boards. The resistance of asbestos boards (asbestos insulation boards, partition boards, flexible sheets, compressed sheets) to corrosion and moisture makes them a popular choice to be recommended as internal roof or wall lining; some particularly hard and thick forms can even be placed in contact with the pigs, and are used alone for pen divisions and gates up to $\frac{1}{2}''$ thick.

Aluminium. As a support for an insulation filling, aluminium is a lightweight lining which only needs a good vapour-seal at the joints to make it a good material for this purpose. As a foil, bonded to kraft-paper, it is also an acceptable lining.

Both these constructions for roofs are more for the specialist manufacturer than the do-it-yourself amateur. They provide a very hygienic surface.

Expanded Plastics. The standard types of lightweight expanded plastic sheets — polystyrenes — are widely used but not always successfully. I would advise their installation as insulation sheets associated with a ventilated air space on the 'cold side' — that is, above the sheets in roof construction and on the outside in walls. They must be protected from the birds, and as their surface is relatively fragile and porous a clean under-lining would appear to be advisable, though at present this is rarely used.

However, the disadvantage of the earlier forms of expanded polystyrene appears to have been overcome in the development of the new very hard extruded forms, such as 'Styrofoam' and these are now widely used as roof linings. They are light enough to make their installation cheap and simple, yet still hard enough to give a clean and hygienic surface.

Another more recent expanded plastic—polyurethane—can be used in three ways. It may be pumped into a cavity forming an insulation material of extremely light weight, rather like 'candy floss' in texture. It may also be sprayed on the underside of the outer cladding, such as asbestos, to form a hard inner insulation lining. Finally, it is now made in sheets like polystyrene but harder and often faced with plastic to give a fine vapour-proof and hygienic surface (e.g., 'Purlboard').

Even more recently polyurethane, which has a better insulation value than polystyrene, has been effectively bonded to external cladding materials such as oil-tempered hardboard, resin-bonded plywood and plastic-coated steel and aluminium sheets.

This technique produces in one structure a hard-wearing well-insulated material that can form the entire roof and walls of piggeries provided the joints are permanently sealed.

The inside surface is normally plastic-coated, but where the walls are in contact with the pigs they may be protected with external cladding materials sufficiently strong to withstand the animals, for example, $\frac{3}{8}''$ fully-compressed asbestos.

Plywoods. 'Exterior grade' phenolic resin-bonded plywoods can be successfully used as interior or exterior lining, but for external use they should have good weathering protection. This may be a liquid bitumen preparation.

Compressed Straw Slabs. A well-known brand of compressed straw slab suitably protected from the weather is often used in the roof as cladding and insulation combined. As this gives a "U" value of over 0·2 and our aim is to have a figure 100 per cent better than this, such a construction can only be recommended if an additional insulation lining and good vapour-seal is placed underneath.

INSULATED CLADDING

Constructions such as those given above answer our requirements well, and are usually maintenance-free. There is, however, another approach which has interesting use as an alternative. Certain materials can be used as an insulation *and* a cladding. An example is a wood-wool slab. Using a 2-inch or 3-inch thick slab of this material a very high degree of insulation is given and when used on the roof, it will only need a weatherproof covering to complete the job. It is probably also advisable to render it on the inside to give a smooth hygienic surface that can be adequately cleaned.

We have actually used this material for the roof of an animal house quite untreated and this has strangely given us no condensation problems at all in spite of the lack of a vapour-seal. The slabs have a fairly open meshwork — they consist of wood-wool fibres bonded together with cement — and appear to 'breathe' at times, taking up water vapour then giving it up later, but apparently without condensation in a well-insulated house. For roofs this is a relatively heavy structure, but for a wall it gives a strong, well-insulated structure which, if cement-rendered inside and out, is satisfactorily hygienic and robust for any form of livestock. A number of piggeries are now being made in this way.

TRADITIONAL METHODS

Where more traditional methods are used for the walls, with

cavity brick or concrete block, many serious errors are still made. Firstly, it should be stressed that the degree of insulation is not too high with these methods. Cavity brickwork is also very expensive and is used less and less for the insulated animal house. Insulated forms of concrete blocks are frequently used, and laid by unskilled labour, but if they are to retain their usefulness, it is vital to protect the outside from rain either by rendering, or treating with some cheaper liquid sealing compound. Ordinary hollow blocks in dense concrete are not insulators and can only be recommended where they are used in conjunction with an inner leaf of lightweight insulation blocks. A wall construction of two layers of blocks, the outer being of dense concrete, the inner of insulation form with an air space between, is a very good construction.

Insulating the Floor

We have not so far mentioned floors, but these are indeed of great importance. Stock are frequently in close contact with the floor and there is usually a need to conserve bedding to the

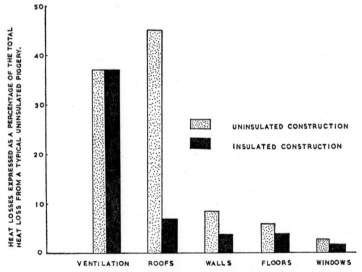

Fig. 11 *Diagram showing where the heat is lost from insulated and uninsulated piggeries.*

54

utmost. The use of an insulation layer is therefore strongly advisable in a building — or that part of a building — where pigs have direct contact with the floor. Further, in all cases where a concrete floor is installed but the use of a thick layer of litter makes insulation unnecessary, a damp-proof course should be used to prevent rising damp. Good floors would, therefore, consist of the following layers, from bottom to top: hardcore, concrete, damp-proof course, insulation layer, cement screed. The damp-proof course can be of polythene, kraft-paper, roofing felt, or a bituminous liquid. Insulation layers can consist of wood-wool slabs, egg trays, hollow tiles, domed concrete blocks, or expanded plastic. Naturally the top screed should be as thin as possible—for pig pens it is usual to make do with a 1½-inch thickness, but care must be taken that the screed is thick enough and well made so there is no risk of collapse.

Much damage is done to pigs by using rough concrete surfaces and the finish must be smooth enough to prevent this, yet not so smooth that the pigs slip. A wooden-float finish is best.

Reflective Insulation

The methods of insulation so far described rely on the dead-air spaces within the material reducing the conductance of heat. There is, however, another very useful, and often cheap way of reducing heat loss by so-called 'reflective insulation'. A bright shiny surface — such as aluminium — will reflect much of the heat back that is radiated towards it. Thus it is extremely useful to line the inside of a house with aluminium foil as this will conserve the radiation emanating from the livestock, heaters or infra-red lamps, or even other parts of the building itself. Reflective insulation is particularly useful as a lining to the inside of the piglets' creep and 'kennels' for fattening pigs.

Reflective insulation on the outside of a building — for example, using an aluminium roof covering and white-painted walls — will materially assist in lowering inside temperatures on hot summer days by reflecting part of the sun's heat.

Insulating the Windows

The controlled-environment animal building is often made without windows. Some may deprecate the cult of rearing stock under completely artificial light, but it does have some real advantages. For in this way one can control the activity of the stock so that problems such as fighting, cannibalism and sheer over-excitability are alleviated. And from the point of view of insulation it has the great advantage that a most serious source of heat loss and condensation is eliminated. Where, however, windows are used in such buildings it is at least essential to use a double thickness of glass with an air-space between. This is not over-expensive and greatly reduces the heat loss. As pigs tend to go towards the light to dung the windows should be in or over the dung passage.

In connection with thermal insulation, it is worth stressing the merits of reducing the air space in a house to reasonable proportions. Buildings still tend to be unnecessarily large, and roofs in particular too lofty. By reducing the cubic area, heat conservation is improved and building costs can be actually reduced. It is also most important to have efficient drainage around the building. It is surprising the number of cases where the floor becomes permanently wet throughout the winter due to inefficient drainage. Such a floor is certain to be cold and any litter on it will absorb a great deal of moisture and any stock may be chilled.

Modern intensive unit on an East Riding farm. It comprises of a weaner house, two farrowing houses and a fattening house.

Fattening house at Middle Aston—a Danish-type layout with screened-off dunging passage and stale-air extraction in walls.

Automatic feeding system in this package-deal fattening house dispenses pellets onto pen floors six times a day; 68 pens each hold 15 pigs to bacon.

This Danish-type fattening house has controlled ventilation providing a temperature of 60°–70°F. Air duct runs down centre of the building. Swinging pen-fronts are in "feed" position. The house has no windows.

This well-ventilated and insulated piggery on a Yorkshire farm has railed pen fronts and central extracting ventilators level with ceiling.

Top picture shows prefabricated, fully-insulated fattening house with low roof giving minimal air space; lower picture shows interior of the building showing stretched-wire pen-fronts.

Above: Fully-insulated farrowing house with 32′ clear span and suspended ceiling. *Left:* View of dunging passage in same building. Farrowing houses with dunging passages must be depopulated periodically.

For added warmth in this double Harper Adams layout the pen divisions reach up to ceiling height, and self-closing double doors are fitted to feeding passages.

Chapter 4

VENTILATION SYSTEMS

THE conventional method of ventilating the piggery is to extract the air from a limited number of chimney-type extractors, whilst the air enters through baffled inlets around the wall. With natural ventilation extraction area should be 5 square inches to each 100 lb liveweight of pigs housed, or with mechanical ventilation a maximum of 20 to 25 cubic feet per minute (cfm) per 100 lb liveweight, depending on the area of the country, the higher levels applying to the South and West and the lower figures to the North. The inlet area should also be related to these figures; to each 1,000 cfm extracted allow 5 square feet of inlet area, or to each pig allow 15 square inches of inlet area at least. This means in practice that the inlet area should be three times the outlet area, and this enables one to be sure that the inlet velocity is low and the likelihood of draughts is minimised.

In planning the ventilation on these lines it is preferable to have one chimney outlet to each section of the house. Thus if there were 100 bacon pigs in a common air space, the chimney outlet would be 1,000 square inches or 33″ × 33″ in size. If mechanical ventilation was desired, the pigs would require a maximum of 5,000–10,000 cfm, depending on the locality. In the base of this chimney could be inserted a 24″ extracting fan running at a maximum of 900 rpm to give 6,000 cfm which

57

would be suitable for the colder areas. The advantage of this system is that it is most flexible, as it is a truly natural and mechanical system combined, and the fan assists or supplements natural flow of air. If the fans or electricity fail, then the natural flow is always capable of carrying on until they are restored.

The diagram (fig. 12) shows the nature of the chimney extractor which is of two-skin construction and fully insulated. Some form of control, such as a hinged or sliding damper or

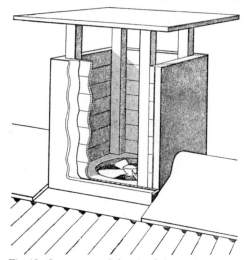

Fig. 12. *Constructional details of chimney extractor.*

butterfly valve, must be placed at the base to regulate the quantity of air passing through the chimney. The chimney may be placed on or just to the side of the ridge. If there is a flat false ceiling the outlet chimney should be extended down to it and this also must be insulated the full way to avoid any risk of condensation and down-draught. This is particularly important when metal or asbestos ducts are used.

The fresh-air inlets, on the other hand, should extend all around the walls and form a nearly continuous line at least 3' 6" above the pigs, but not less than 1' below the eaves; if either of these limits is exceeded there is danger of draughts on the pigs. The detailed design of the air inlet also needs great care and a suitable form for normal sites is shown in the accompanying illustration (fig. 13). The main features are the bottom-hinged, inward-opening control flap on the inside and the hood on the outside. The control flap on the inside must

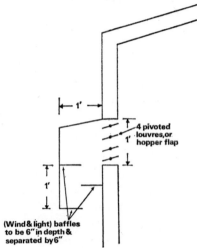

|← 1' →|

4 pivoted
louvres,or
1' hopper flap

1'

(Wind & light) baffles
to be 6" in depth &
separated by 6"

Fig. 13. *Detailed design of air inlet.*

be controllable in stages; a common mistake is to make it so that it can only be fully open or closed tight shut, which greatly limits the flexibility of the system; in exposed positions the outer hood also needs especial care to prevent the wind using it to funnel the air into the piggery. One method of dealing with this is to place within the hood wind and light traps, as commonly used in controlled-environment poultry houses. It is done by placing two or three boards along the length of the hood, each being half the width of the hood and lying some 6" apart and fixed on opposite sides. The alternative arrangement is to open at least part of the top of the hood to allow the wind an escape route or to place a deflecting board a few inches below the botton of the hood. An even better arrangement is shown for the inlet flap with several pivotted louvres.

It is desirable that air inlets are constructed of maintenance-free materials, such as treated wood, plywood, or asbestos, and the inlet flap itself should be of two-skin insulated construction to prevent condensation and deterioration.

FAN INSTALLATIONS

Fans are powerful tools and must give a high extraction rate to cope with the maximum summer requirements and when the piggery is fully stocked. Much reduced rates are needed when the weather is cold, the house only partially full, or the pigs are well under their maximum weight. For example, a full-grown baconer in summer will need up to 100 cfm. In the coldest winter weather about one-sixth of this would be needed for this weight of pig—say, 15 cfm, but a weaner in the same house in winter would only need about 4 cfm, that is, one twenty-fifth of the maximum rate. Fortunately it is possible to utilise special agricultural fans that can be reduced in speed to some 10 per cent of the maximum and the fans may be switched off by thermostat or gradually reduced in speed by either a motorised thermostat or a thermistor and solid state control, depending on the temperature of the house.

With multi-fan systems as used in most piggeries, there are two control arrangements that may be used. One method is to allow about one in three of the fans to by-pass thermostatic control and to be only on manual operation. This is to make sure that at least some ventilation proceeds at all times. With this arrangement of having some fans manually, and some thermostatically controlled, the latter should be fitted with automatic anti-back draught flaps to prevent down-draught when the latter are 'off' and the former are 'on'.

Thermostats should be sited fairly close to the inlets and in the direct line of the air-flow; a distance of 6–10 feet is advised to give an adequately quick response. Sometimes speed controls are augmented by timer controls. The fan-timer cycle may be five minutes. Within this cycle the fans can be brought in from 10 per cent of the time upwards to 100 per cent. This is quite a useful augmenter to a speed control, but should never be used as a *substitute* for one.

To some extent and especially in large buildings the motorised thermostat or sensitive thermistor control has replaced the on-off thermostat. The great advantages of these methods are that they *gradually* change the fan extraction capacity with pre-set and variable minima according to the basic needs of the pigs. All fans can be regulated together and the use of a pre-set minimum rate which the stock-man can vary as he wishes ensures that there is always sufficient ventilation, whatever the weather and whatever the age or weight of pigs.

Examples of suitable fans and their capacities are shown in the table below. By and large it is more economical to use larger fans at high-speed maxima, but such units can only be used in the larger or wide-span buildings unless a certain amount of ducting is used to spread the collecting area over a wider area.

Alternative Systems

The arrangement described above is suitable for many forms of piggeries, and more particularly for those of totally-enclosed design with central dunging passages. There are, however, a number of interesting alternatives which are gaining in popularity and rely on entirely mechanical measures.

(a) *Extraction via dung passage walls*

Because the majority of the most unpleasant effusions in the piggery arise from the dunging area it has become popular in those piggeries with dunging passages alongside the outer

Suitable Fans for Pig-house Ventilation

Fan	Size r.p.m.	Approximate maximum air delivery	Number of pigs
12″	1,400	1,000	Up to 20 baconers, or 25 porkers, or 5 sows and litters
15″	1,400	2,500	Up to 50 baconers, or 60 porkers, or 12 sows and litters
18″	900	2,500	Up to 50 baconers, or 60 porkers, or 12 sows and litters
18″	1,400	3,600	Up to 72 baconers, or 90 porkers, or 18 sows and litters
24″	700	4,500	Up to 90 baconers, or 110 porkers, or 22 sows and litters
24″	900	6,000	Up to 120 baconers, or 150 porkers, or 30 sows and litters

Maximum rates that should be used are 20–50 cfm per 100 lb liveweight. In warmer areas of the country it is wise to provide as much as twice the maximum ventilation rates given.

walls of the piggery to extract the stale air from there and bring the fresh air into the piggery at or near the ridge. This is a particularly suitable system for the Danish-type piggery with screened-off dunging passage, and is also strongly advised where there are slatted floors on the outside with drainage channel underneath since noxious gases given off and rising from it may cause illness in the pigs and be of serious danger to the pigman. Indeed in all cases where slatted floors are used the extraction point should be over the slats. Where fans are used on the outside walls in exposed sites, it is advisable to place small upright chimneys outside to prevent wind affecting their correct action.

Multiple-fan units are used in the dung-passage walls and the air is brought into the piggery either through chimney-type inlets on the ridge or through a continuously open ridge. The detailed design of the inlets is of supreme importance in the correct functioning of such a system. Where a chimney inlet is used, the open area at the top immediately under the cap should have a cruciform-shaped divider to deflect the wind and fresh air down the chimney. At the base of the chimney

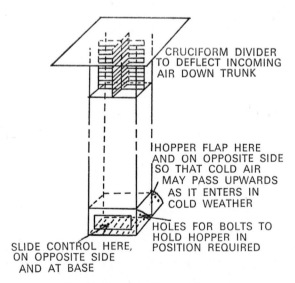

CRUCIFORM DIVIDER TO DEFLECT INCOMING AIR DOWN TRUNK

HOPPER FLAP HERE AND ON OPPOSITE SIDE SO THAT COLD AIR MAY PASS UPWARDS AS IT ENTERS IN COLD WEATHER

HOLES FOR BOLTS TO HOLD HOPPER IN POSITION REQUIRED

SLIDE CONTROL HERE, ON OPPOSITE SIDE AND AT BASE

Fig. 14. Reverse-acting air inlet trunk.

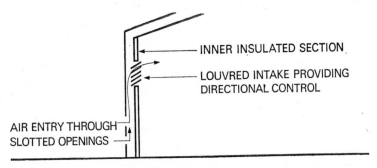

Fig 15. *Air intake arrangement controlling wind and light, and giving good directional regulation.*

there must be control flaps to give directional flow of air. This is done by blocking the base of the chimney and allowing the air to enter the piggery through control flaps in the side. Using these flaps, the air may be directed away from the pigs in the colder weather and towards the pigs in hot summer conditions.

A different inlet arrangement altogether with this system is to use a continuous open ridge but control the amount and direction of the air entering by centrally-hinged flaps controlled remotely at the ends of the house.

(b) *Extraction via drainage system*

A fairly elaborate arrangement, which has been used successfully in a number of piggeries, is to utilise special fans of powerful ability to pull the required quantities of foul air out of the piggery via the drainage system. Aerofoil fans are those normally used. Air enters by the ceiling or along the top of the wall. Though a fairly costly arrangement, it succeeds in producing the pleasantest environment of all systems for pigs and pigman, but care must be taken with the design that the extracting points cannot be blocked with dung. It has been used successfully in association with slatted floors, forming an extraction duct of the drainage channel underneath.

(c) *Plenum or 'pressurised' ventilation*

There are several cases where air may, with advantage, be forced into the piggery by impeller fans. Such instances are the Danish-type piggery with fans placed in or near the ridge and

63

above the centre passage, and those piggeries with outside yards where, by forcing the air into the building through the wall on the opposite side to the yard, the likelihood of draughts is reduced since air should pass out of the pen into the yard. Strong winds may always overcome the effect of the fans, so care with construction of the baffle doors is still required. This arrangement is also used successfully with slatted floors, the foul air passing out from under the slats.

The plenum system may quite simply be achieved by installing a fan in the type of ridge chimney described in section (a). Alternatively, it may be done by placing fans in the gable ends and using an insulated duct running along under the ridge in the angle between the wall and the ceiling or above the false ceiling where present. With such an arrangement great care must be taken to prevent down-draught on the pigs by placing deflectors below the air exits, or directing air towards a wall or passage. Alternatively the base of the duct may be constructed of peg-board or a slotted hardboard to allow the air to diffuse out gradually. The outlets from the duct must keep the air velocity low—approximately 200 feet per minute—and be fully controllable, usually by sliding dampers in the duct. The fans must also be speed- and thermostatically-controlled. Ordinary propeller fans are not always suitable for this application, and aerofoil or other special forms may be preferable.

An additional way of achieving pressurised ventilation is to blow the air into a polythene duct with numerous holes pierced along its length. The precise arrangement for these holes usually requires specialist advice to ensure a uniform diffusion of air; in addition the holes must be placed to avoid a down draught on the pigs. Systems using polythene ducts are most useful in conversions of buildings where it may be difficult to find the space for a more conventional arrangement.

Superficially, the cost of a plenum ventilation arrangement may appear to be high but in practice it can usually be installed for either the same or only slightly more than conventional mechanical arrangements. If well designed, the plenum system always has the advantage of providing a system completely in the pigman's control. Provision should be made with plenum systems for either an emergency generator if power fails or hatches in the wall that can be opened to allow natural ventilation until power is restored.

(d) *The re-circulation system of ventilation*

Many otherwise good systems of mechanical ventilation fail because of their complexity. Farm labour does not have the time nor the skill for servicing electrical apparatus and the result is that breakdowns occur more than they should, and the effect on the stock may be serious.

The simplest arrangement of all with ventilation and involving a minimum amount of apparatus is the 're-circulation' system which may achieve good ventilation using only one fan which runs at a fixed speed (fig. 16). The fan is mounted three to four feet from the end of a duct running above or below the ceiling. The duct is of two-skin insulated construction and contains outlets at the base which should be placed centrally over feed or service passages, to avoid down-draughts on the pigs. From the fan the duct is continued to the gable end in the same way to open to the outside through louvres but there is one major difference. At the base of the duct is a hinged shutter.

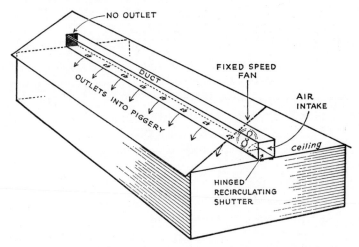

Fig. 16. *Showing the re-circulation system of ventilation described above.*

65

When the shutter is fully closed (see diagram on page 65), the fan will pull all the air from outside to distribute entirely fresh air into the piggery. When, however, the weather becomes cold or there is less weight of stock in the building, the shutter is moved back progressively and as this is done less fresh air comes in but more of the warmed air within the building is re-circulated. Stale air passes out at the side of the building in small outlets 2′ 6″ from the floor. If desired, the re-circulating shutter may be mechanically- and thermostatically-operated, but a good stockman will be able to give superior control to both of these. If the building is more than 50′ long, fans should be placed at both ends. It cannot be used in this way for buildings divided into individual compartments unless the ducts draw in the air from the sides of the house, one duct serving one compartment. Maximum rates that should be used are 25–50 cfm per 100 lb liveweight.

The smaller or slow-speed fans can be used without ducts; the larger or high-speed units will generally require the assistance of simple ducting in hardboard or plywood to give a better distribution of the air.

VENTILATION OF COVERED YARDS

The current tendency towards housing pigs in kennels in covered yards, often of wide spans up to 60′ or 70′, has shown the need for much better attention to their ventilation. For some reason that is hard to explain there has been a tacit assumption that pig housing of this type does not need any special attention paid to ventilation. The serious outbreaks of pneumonia that often occur under these circumstances are very definite evidence of the danger of this belief.

The following general policy is suggested for the ventilation of yards:

(1) In all cases the ridge should have a ventilated opening. 1′ wide for yards up to 40′ width and 2′ wide thereafter. The diagram (page 67) shows the way in which this can be achieved.

(2) At the gable ends, and along the sides above a height of

4' use as much spaced-boarding as possible. Spaced-boards are made with either 4" boards leaving a gap of $\frac{3}{4}$" between each board, or 6" boards with a gap of 1". Or on sheltered sites the gaps may be increased by $\frac{1}{4}$" in both cases. It is also a great help in warm weather if provision can be made to hinge or slide open large sections of the boarding at the end or sides to get uninterrupted cross-flow.

(3) Provision should always be made to allow air to escape from the top of the kennels, or a ventilating flap may be placed in one of the sides of the building. This ventilation must be controllable, but it can be done in simple ways. For example, box-type chimneys can pass through the top with a sliding shutter to control them. Alternatively, slatted boarding can be used as the ceiling, with straw on top, which can be partially removed to give ventilation according to the season.

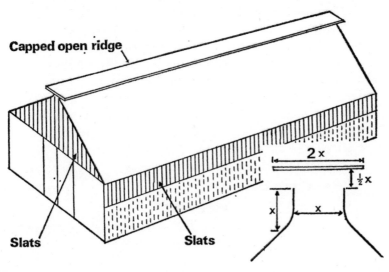

Fig. 17

Chapter 5

CLIMATIC CONDITIONS IN PIGGERIES

AS we have shown previously, pigs will grow most economically and maintain the best health only if the climatic conditions in their house are favourable. As our knowledge of the climate (or, strictly speaking, 'micro-climates') most suitable for pigs increases, the farmer will want to know whether the buildings do provide conditions within the optimum range; if not he can then take measures to correct defects in housing or management or both. The values that can be recorded on the farm, without difficulty, include temperatures, humidities and air currents or 'draughts'. It may be possible on occasions to measure also the air change or ventilation rate in a building.

Temperature

There is no doubt that temperature data are of the most immediate practical use. For this, a continuous-recording instrument, such as a thermograph, is of the most use to the farmer. This is essentially a bimetallic strip that contracts and expands in direct relation to the air temperature. To this is geared a pen which leaves an inked continuous record on a revolving chart worked by clock-work which may revolve once every 24 hours or (preferably) seven days. At the end of this period the clock is removed, a new chart replaces the completed one, and the pen is refilled with special slow-drying ink. That is almost all the attention it needs apart from an occasional recalibration of the thermograph against an ordinary mercury thermometer with the National Physical Laboratory mark on it. A thermograph can be bought for under £30 or hired for a small fee from the Meteorological Section of the Defence Ministry.

Maximum and Minimum Thermometer

A second instrument that has considerable use for measuring air temperatures is the maximum and minimum thermometer. Movement of the alcohol and mercury in this U-shaped instrument pushes small metal pointers to the lowest and highest temperatures experienced since the pointers were last reset, this being done with a small magnet which brings them back to the current air temperature. If the readings are taken regularly twice daily (for instance, at feeding times) a useful record of the range of temperature conditions in the house can be built up.

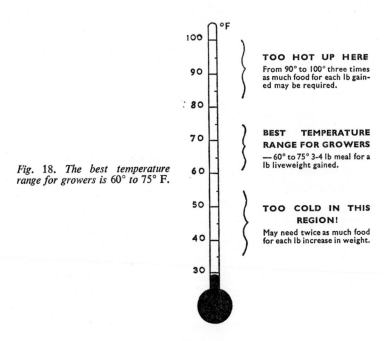

°F

TOO HOT UP HERE
From 90° to 100° three times as much food for each lb gained may be required.

BEST TEMPERATURE RANGE FOR GROWERS
— 60° to 75° 3-4 lb meal for a lb liveweight gained.

TOO COLD IN THIS REGION!
May need twice as much food for each lb increase in weight.

Fig. 18. *The best temperature range for growers is 60° to 75° F.*

The cost of this handy instrument is only about £1, but it has quite obvious limitations compared with a continuous recorder. For instance, any fluctuations within the extreme range since the previous readings were taken are lost; also one gains no idea when the changes occurred or how long they lasted.

Either of these instruments should be placed near the stock rather than somewhere which is handy for quick reading but does not show the environment experienced by the animals. This means, in many cases, that the instrument must be protected from damage by the pigs by a screen; this is perfectly satisfactory provided it allows free circulation of air through it.

As well as recording the air temperature in one or two selected sites in a house, it is advisable to take 'spot' measurements from time to time of gradients of temperature from end to end, side to side and floor to roof. If these are made in a number of positions, a very useful indication is given of the presence of draughts, stale air pockets and cold and hot spots. Also, it may be found useful to take surface temperature readings. These will give some indication of roof and wall insulation and also litter temperatures. A flat-bulbed thermometer (preferably surrounded on the open side with a shiny metal sheath) and fixed on with modelling clay is all that is needed for these measurements.

Humidity

The interested farmer or stockman sometimes records the humidity of the air, and almost always he uses Mason's wet and dry bulb hygrometer. This consists of two thermometers suspended side by side, the bulb of one being surrounded by a wet jacket. So long as the air is not completely saturated with water vapour, the reading of the wet bulb will be lower than that of the dry one as water vapour will evaporate from the jacket and cool it. By using tables, the observer can find the *relative* or *absolute* humidity of the air from the dry and wet bulb readings. The relative humidity is generally the figure required — usually expressed as the percentage of water vapour in the air, 100 per cent being saturation point.

However, Mason's hygrometer is really notoriously unreliable as little air may pass over the bulbs when it is hung in a building, and so vapour may not evaporate properly. A modified form of this instrument, known as the whirling hygrometer should therefore be used; in this the two thermometers are arranged as in Mason's but are mounted so that they can be rotated in the air to cause a current of air — representative of conditions in the building — to flow over them.

Very useful instruments also exist that enable direct readings of relative humidity to be obtained. The hygrograph, which is the companion to the thermograph, gives a 24-hour or weekly reading of the relative humidity on a drum. The basis of this instrument is a bundle of specially treated human hairs which contract or expand according to the relative humidity of the air; simpler hygrometers on this pattern but merely giving 'spot' readings or relative humidity are now manufactured to give a 1-2 per cent accuracy. This is quite sufficient for our purposes, and although these hygrometers are more expensive their simplicity makes them attractive. The same rules for their placing should be followed as for thermometers.

Draughts and Ventilation Rates

Instrumental measurements of air velocities (draughts) are rarely made, but in some cases they can be most useful. Smoke from cigarettes or pipes has a certain limited use as an indication of the way the air is flowing, but it is too transient to give a clear picture. A much more satisfactory indication is given by the fumes of titanic chloride, a chemical which fumes copiously on contact with air. Titanic chloride has the advantage that it can be used carefully when the house is fully stocked as it is not prejudicial to health in small quantities. Alternatively, small smoke candles may be purchased to do the same job.

Chapter 6

DISINFECTION ROUTINE AND DISEASE CONTROL

EVERY pig farmer recognises that pigs tend to 'do' better in new buildings than old. This is because, so we believe, there is a gradual 'build-up' of disease-causing organisms in and around the structure of the house. This need not be accepted as inevitable in this day and age when the causes of infectious disease are generally understood and the means of preventing them within our reach. The ideal basis for dealing with the problem of build-up is to arrange all accommodation so that it is not too large to be periodically emptied; the necessity for this, according to all practical experience, is in inverse proportion to the age of the pigs housed. Thus in farrowing, rearing or early-weaning accommodation it is absolutely essential, whilst the need for it is less but not absent in accommodation for fatteners and adults. Also, we must never forget that the reservoirs of infection are not only in the structure of the buildings, but in the pigs inside them. When we empty a building, or 'depopulate' as it is usually described, the process must be complete to be effective and must include every pig in the house. A period of two or three weeks should be allowed for the emptying and disinfection, as in practical terms this is usually required to give the assurance of absolute cleanliness.

Gas heater of a type that can be used for warming piglet creeps.

Lid to creep in farrowing house supports infra-red lamp and reduces heat loss.

Simple protection for the piglets, using farrowing rails.

Convertible farrowing/rearing pen.

Simple crate used in farrowing house; note cement-rendered wall and metal fittings—both aids to hygiene.

Farrowing crates in a Pitmillan type house, with removable dwarf pen surrounds.

Solari farrowing pen.

This solid-wall farrowing pen has a single low farrowing rail. There is a lidded creep in the feeding pass.

Interior of rearing house.

Slats in this farrowing pen extend along one side of the crate, where they are more likely to dung

After the piggery has been emptied of all stock the following procedure may be applied:

1. *Removal of Muck and Cleansing*

This is probably the most important part of the operation as pathogenic organisms will not generally live for long in a well-built house in the absence of protective organic matter. All muck should be taken out and placed as far away from the premises as possible. This is very important. The subsequent cleansing may then be carried out in several ways. Some of the alternatives are:

(a) By water sprayed under pressure. This is an effective and popular way in agriculture, being cheap and practicable.

(b) Steam cleansing. This is effective both for cleaning and disinfection, using a suitable steam generator. The equipment is expensive, however, and the operation laborious.

(c) Soak and scrub with hot water containing a detergent or four per cent washing soda.* Soaking may be done in cold water followed by scrubbing but this represents the most laborious method of all.

(* A double handful in a bucket.)

2. *Disinfection*

When cleansing is completed, the equipment should then be washed, sprayed or soaked in a disinfectant approved by the Ministry of Agriculture and used at recommended dilutions. It is worthwhile having suitable tanks or baths built for this. The process of detergent cleaning and disinfection may be combined by using a disinfectant with detergent properties and this in many cases will be the method of choice.

3. *Washing*

After 24–48 hours the disinfectant should be washed down with clean water, paying particular attention to any of the equipment to which the pig has close contact.

N.B. Certain heating equipment, such as infra-red lamps or gas heaters, need special care to prevent damage to their mechanism. A suitable way to deal with them is clean off all dust with a vacuum cleaner, scrape off any adherent matter, and place in a room for formaldehyde fumigation.

Alternative procedures exist that make use of the valuable new knowledge we have on the benefits of 'aerosol' fumigation, that is, the dispersal of gases or vapours of very small size. In applying these, the procedure is at the start similar to the first-mentioned —that is, remove all pigs and clean the building as thoroughly as possible. Then after cleaning, use an aerosol of small particle size for spraying formalin, at the rate of 2 fl oz per 100 cu ft of air space for the smallest buildings, 1 fl oz per 150 cu ft for buildings up to 12,000 cu ft and 1 fl oz per 300 cu ft for the argest buildings. Dilute the formalin in water before spraying (1:1 dilution) or disperse the water first, as high humidity increases the effectiveness of the formalin. It is also advantageous to warm the house first to 60°–70°F if possible. Leave the building closed for 24 to 48 hours.

Follow by spraying with an aerosol of large particle size comprised of six parts 4% chloroxylenol mixture and four parts triethylene glycol at a concentration of 4 fl oz per 100 square feet of floor space. Alternatively, many farmers prefer the convenience of using one of the excellent proprietary disinfectants that may be used in aerosol form. This procedure acts as a residual disinfectant.

It is vital to seal up the building as thoroughly as possible. A good procedure is to tape round doors, windows and ventilators.

Formaldehyde Gas Dispersal

In the absence of an aerosol machine to dispense the formalin, the formaldehyde gas may be dispersed in one of two other ways. The easier is to purchase formaldehyde fumigating lamps. At a cost of approximately 25p, 1,000 cu ft can be dealt with, the lamps being supplied with small burners and the correct quantity of solid paraformaldehyde. An even cheaper way is to generate the formaldehyde gas by the addition of liquid formalin to potassium permanganate crystals. Two fluid ounces of formalin per 100 cu ft of air space are placed in metal containers

to deal with buildings up to 12,000 cu ft, and for larger buildings up to 24,000 cu ft one fluid ounce per 100 cu ft. The
formalin is always added to the permanganate—three parts of
formalin mixed to two parts of permanganate. The permanganate should be placed in a deep-sided vessel and as the com-
pounds react with violence the operator must withdraw from
the room quickly to prevent any ill-effects. The receptacles should
not be placed near any combustible materials as there is a considerable amount of heat evolved and fire can result. The safest
procedure for the operator is to use a gas mask. It is unwise
to place more than 1 lb of permanganate crystals in any one
container and the vessel sides should not be less than three times
the depth of the chemical.

Chapter 7

FARROWING ACCOMMODATION

THE pig farmer should be ever mindful that there is an appalling mortality in piglets before weaning; an average in excess of 20 per cent of those born alive. He should also know that the majority of these losses are due, not to disease, but to bad management, housing playing an important, if not principal, part in this. Surveys have shown that over 50 per cent of the piglets that die perish due to chilling and crushing; also much of the disease that occurs may be induced by the stress of rearing them under unsatisfactory conditions. The yardsticks for farrowing quarters may be said to be protection, warmth and hygiene. Full details of the climatic requirements of the sow and piglets have been given in chapter 2.

The sow is a clumsy mother, the worst on the farm; some breeds are worse than others but the breeds which are certainly good mothers, such as the Saddleback, have decreased in importance, whilst the traditionally more difficult ones—the Large White and Landrace—have pushed most other breeds into the background. Some form of protective crate for the sow can be considered therefore essential to prevent her clumsy movements crushing the piglets. At the same time, the piglets must be encouraged to spend their resting periods away from the immediate proximity of the sow and this is best done by providing nests close to, or as part of the crate itself. Ideally

76

artificial heat and light sources should be placed above the nest to attract and warm the piglets.

It must be borne in mind that the pigman has to keep a watchful eye on the sow and piglets over the farrowing period without worrying or unnecessarily interfering in any way. In other words, it is as important to give him good facilities as it is for the piglets! All too often this is quite forgotten. We can say the piglets need a draught-proof nest temperature of 70°–80°F and the sow and pigman, 50°F minimum.

For these various reasons, the totally-enclosed farrowing house of specialised design is gaining in popularity. It is a trend that can be fully supported. Nevertheless, every farmer should be aware of the fatal error of making the house too big. Sows and piglets need hygienic surroundings, which means that an essential requirement is periodic depopulation, fumigation and disinfection of the building. Also, sows need quietness and they get this much better in the smaller building. We aim, therefore, to have a unit of a maximum of 16–20 pens within a building, and preferably less. But the limiting factor is that it must be small enough to be emptied of all stock regularly.

The Farrowing Crate Unit

In its simplest conception a farrowing crate consists of a pair of three parallel rails. The top rails are set 1′ 9″ apart, as also are the centre rails. The bottom rails are 2′ 6″ to 2′ 8″ apart, depending on the type of sow, and to allow adequate room when she lies down. The bottom row are a minimum of 10″ from the floor and the second row is 1′ above the bottom row and 1′ below the top. Thus the crate has a total height of 2′ 10″ from the floor. It can be constructed of 1″ bore tubing which may be fixed into concrete blocks at the front and back of the crate. There are escape nests on each side, a minimum of 1′ 9″ wide. Thus the total width of the crate is 5′ 3″ to 5′ 5″, or with dividing walls approximately 6′. The walls at the outside of the nests and at the front of the crate can be solid, and there is a gate at the back for access by the sow and the attendant. On the inside of this gate there should be a semi-circular metal bar 10″ from the

77

base and extending 9″ inside the gate. This will prevent the sow backing right up against the gate and crushing the piglets.

It is desirable to have a cover of plywood, hardboard or asbestos sheet over each nest, and on each side there should be a heat source—usually an electric infra-red lamp. Great care must be taken with the falls in the floor to make sure that any water or urine runs towards the back and away from the crate and creep; this may seem obvious but it is surprising how frequently the falls are incorrectly made and lead to muck accumulating within the crate. The top on the nest is important to reduce floor draught due to high-speed convection currents induced by the infra-red lamp.

ALTERNATIVE CRATES

The type of crate described, made of metal and block or brick, is probably the simplest, but alternatives are used—crates with timber, plywood and brick and concrete sides. The essential dimensions, however, are the same. As sows vary in size, the gate at the back of the crate is sometimes made to fit in a number of runners situated at distances, measuring from the front 7′ 6″, 7′ 3″ and 7′. Another refinement is to have several metal bars across the top of the crate to prevent the sow rearing up. Though this seldom happens in a small farrowing unit where the sows remain quiet, it is sometimes a problem in a large unit where activity in the house is greater. This fitting is best left portable so that the bars can be used as and when required.

The type of unit in which the crate can be fitted varies tremendously and there are several perfectly acceptable alternatives. As the sow may not be in a crate for more than 14 days, it is not considered important for the sow to have exercise. Indeed, many of the most successful units consist of a row of crates, single or double-sided, with a feeding passage in the front and a small passage for cleaning and movement of the pigs at the back. Cleaning, unfortunately, generally has to be done manually and it is a difficult operation to mechanise efficiently. To ensure easy access to the creep, to attend to the piglets, the walls adjoining this portion need be only 1' 9"–2' high to retain the piglets. A trough and, if desired, a separate water-bowl may be placed in the front of the crate. With this design it is best to use rails in the front, to assist feeding, or it may be

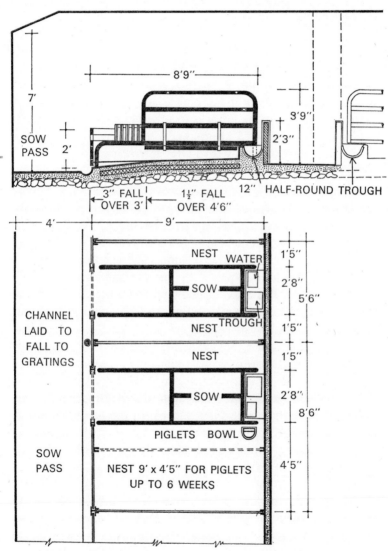

Fig 19. Plan of Pitmillan-type farrowing pen, fitted with Ritchie crates.

79

fitted quite successfully, but more expensively, with a swinging trough-front.

With these units the cross-sectional size of the building can be 16' with a single-sided unit (crate and trough 9', passages each 3' 6"), or with a double-sided unit 28', both being standard sizes in the building industry.

An extremely popular form of crate is the Ritchie farrowing crate in Pitmillan-type pen (see fig. 19). This is a fixed crate (allowing the sow no exercise) with nest or creep area on each side with low walls allowing easy access to the piglets. The nest is 1' 5" only each side if the piglets are removed early before creep feeding, or if the latter is to take place the creep on one side is increased to 4' 5". Sows and litters may be kept in this accommodation up to 6 weeks after farrowing.

BUNKER DESIGN

Some farmers believe that it is important to provide exercise for the sow; the crate can still be used in this form in the 'Bunker-type' arrangement, devised by Mr L. A. Bunker of Dunstable. Here the crates are placed alternately head to tail and there is an exercise area behind each extending over the width of two crates (see page 82). The exercise area is 5' wide and 12' long; in this area are the feed and water troughs, and the sow will probably be allowed out to feed and exercise twice a day for around half an hour.

This design is rather more expensive than the one previously described, a single row of crates only being accommodated in a building 18' wide. It also means 5' wide doors along the exercise and feeding area, and a common passageway for cleaning. Whether the extra expense is really worth the luxury of allowing the sow exercise for the short period during which she is in the crate is rather doubtful, but taken all in all this may be said to be a most satisfactory system for the large enterprise, provided it is divided into smaller units.

The question of disease is always very much in mind with farrowing units, and a few farmers prefer to farrow in non-specialised buildings. For these, the crate can be a portable structure of metal or timber and can be placed in a loose-box

PIGHOUSE PLANS AND FITTINGS

The plans and diagrams in this "blueprint" section are referred to in the appropriate places in the text. They cover the main types of housing for different classes of pigs and also include shelters and useful fittings and equipment for piggeries.

BUNKER FARROWING PEN

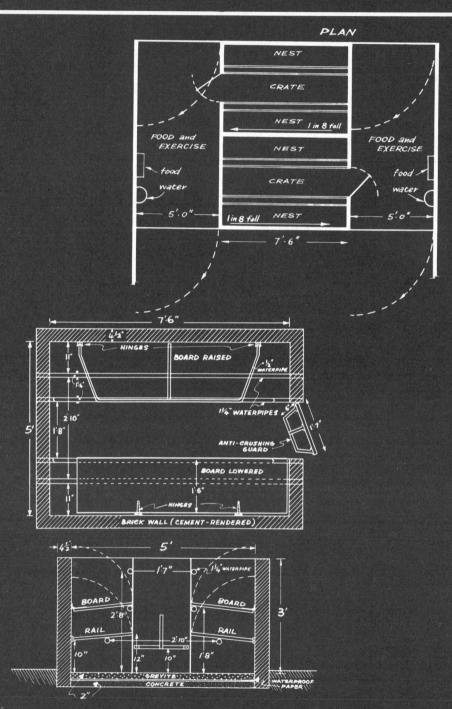

MINIMUM SIZE
REARING PEN

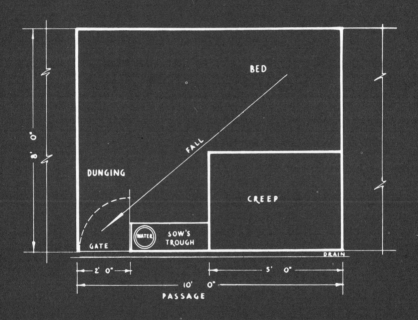

BED

FALL

DUNGING

CREEP

GATE

WATER SOW'S TROUGH

DRAIN

8' 0"

2' 0"

5' 0"

10' 0"

PASSAGE

SLATTED-FLOORED FARROWING CRATE

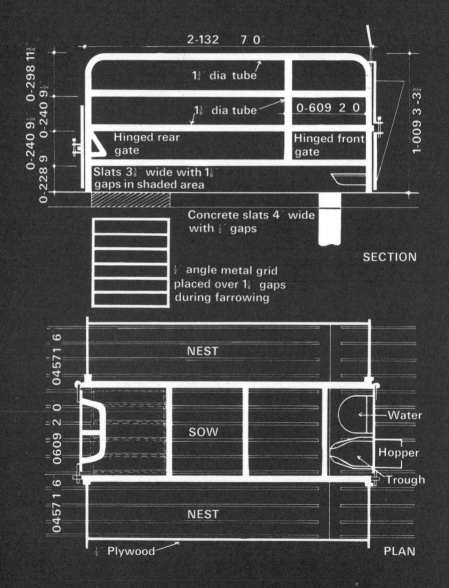

2-132   7 0″

1¾″ dia tube

1⅛″ dia tube

0-609 2 0″

Hinged rear gate

Hinged front gate

Slats 3⅜″ wide with 1⅛″ gaps in shaded area

Concrete slats 4″ wide with ½″ gaps

½″ angle metal grid placed over 1⅛″ gaps during farrowing

0-298 11¾″

0-240 9½″

0-240 9½″

0-228 9″

0-240 9

1-009 3 -3¾″

SECTION

0457 1 6

0609 2 0

0457 1 6

NEST

SOW

NEST

Water

Hopper

Trough

½″ Plywood

PLAN

REARING HOUSE: Muir of Pert Farms

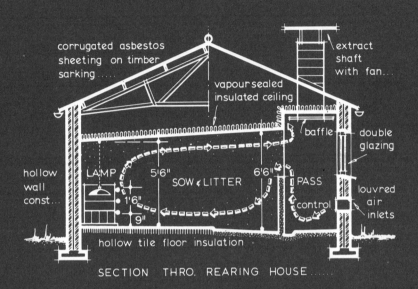

corrugated asbestos sheeting on timber sarking .....

vapour sealed insulated ceiling

extract shaft with fan...

baffle

double glazing

hollow wall const...

LAMP

5'6"

SOW & LITTER

6'6"

PASS

control

louvred air inlets

1'6"

9"

hollow tile floor insulation

SECTION THRO. REARING HOUSE......

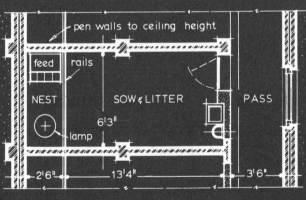

pen walls to ceiling height

feed rails

NEST

SOW & LITTER

PASS

lamp

6'3"

2'6"

13'4"

3'6"

PLAN OF PEN LAYOUT ......

EARLY WEANING PEN

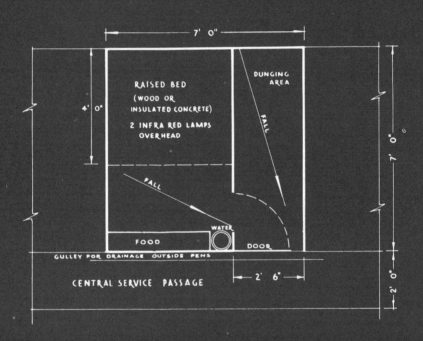

7' 0"

RAISED BED
(WOOD OR
INSULATED CONCRETE)

2 INFRA RED LAMPS
OVERHEAD

4' 0"

DUNGING
AREA

FALL

FALL

WATER

FOOD

DOOR

GULLEY FOR DRAINAGE OUTSIDE PENS

CENTRAL SERVICE PASSAGE

7' 0"

2' 0"

2' 6"

SOW STALLS

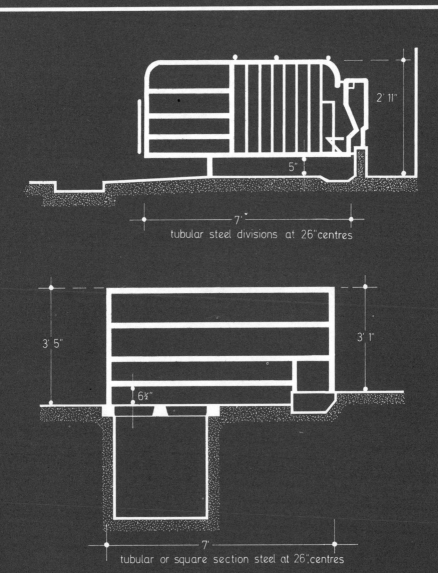

2' 11"

5"

7'

tubular steel divisions at 26"centres

3' 5"

3' 1"

6½"

7'

tubular or square section steel at 26"centres

Top drawing shows stall with solid floor and outside gully. Bottom drawing shows stall with slats and slurry channel.

CRAIBSTONE ARK

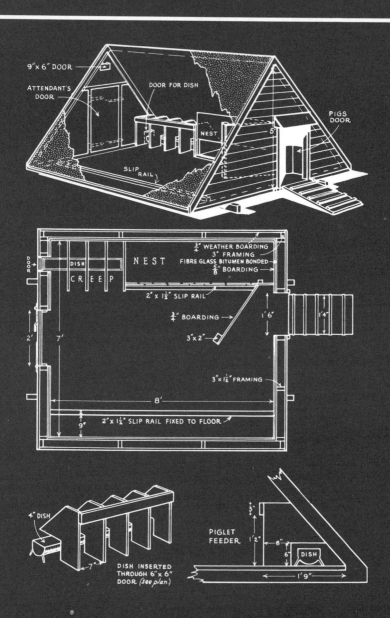

9" x 6" DOOR

ATTENDANT'S DOOR

DOOR FOR DISH

NEST

PIGS DOOR

SLIP RAIL

6"

¾" WEATHER BOARDING
3" FRAMING
FIBRE GLASS BITUMEN BONDED
⅝" BOARDING

DOOR

DISH

NEST

CREEP

2' x 1½" SLIP RAIL

¾" BOARDING

3" x 2"

1' 6"

1' 4"

3" x 1½" FRAMING

2'

7'

8'

9"

2" x 1½" SLIP RAIL FIXED TO FLOOR

4" DISH

7"

DISH INSERTED
THROUGH 6" x 6"
DOOR. (see plan.)

PIGLET FEEDER.

1' 2"

8"

6"

DISH

1' 9"

TOTALLY ENCLOSED DANISH-TYPE PIGGERY

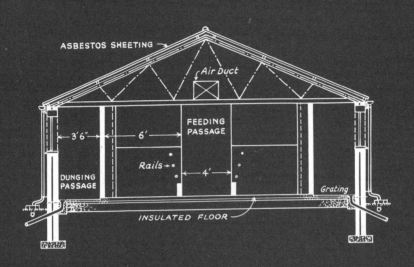

ASBESTOS SHEETING

Air Duct

FEEDING PASSAGE

3' 6"

6'

Rails →

4'

DUNGING PASSAGE

Grating

INSULATED FLOOR

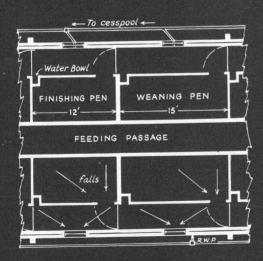

← To cesspool ←

Water Bowl

FINISHING PEN
12'

WEANING PEN
15'

FEEDING PASSAGE

falls

R.W.P.

DANISH-TYPE FATTENING HOUSE
WITH SCREENED-OFF DUNGING PASSAGE

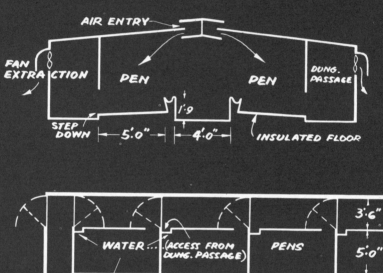

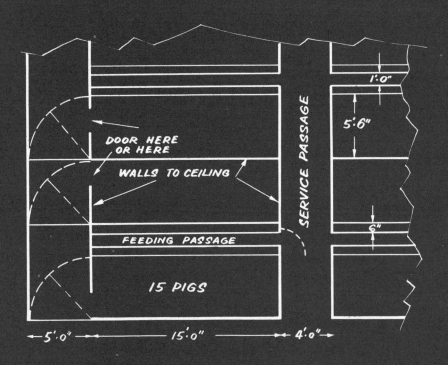

DOOR HERE
OR HERE

WALLS TO CEILING

FEEDING PASSAGE

SERVICE PASSAGE

15 PIGS

1'·0"

5'·6"

6"

5'·0"

15'·0"

4'·0"

Built as single — or double-sided unit.

HOUSE WITH CENTRE DUNGING PASSAGE WITH SLATTED FLOOR

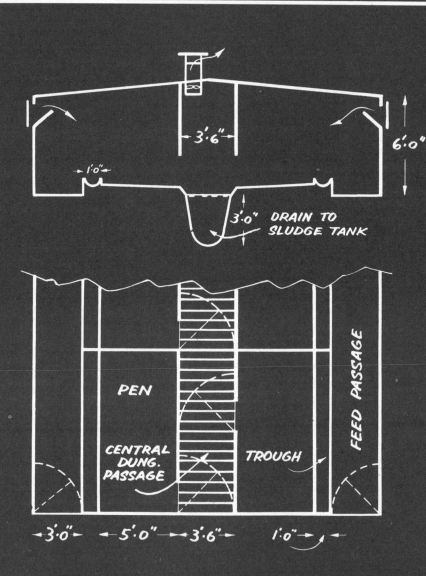

6'.0"

3'.6"

1'.0"

3'.0" DRAIN TO SLUDGE TANK

PEN

CENTRAL DUNG. PASSAGE

TROUGH

FEED PASSAGE

3'.0" 5'.0" 3'.6" 1'.0"

HOUSE FOR FLOOR FEEDING WITH
CENTRE SLATTED DUNGING PASSAGE

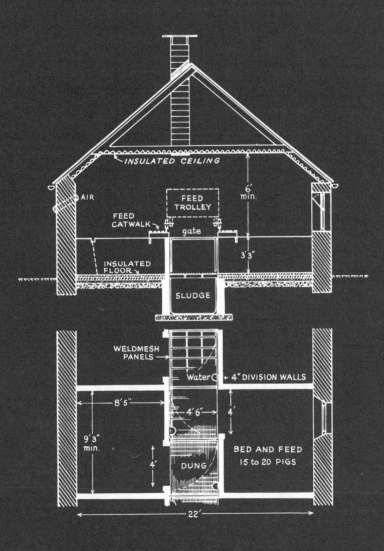

This house has raised feed catwalks overhanging pens on either side of dunging passage, and a feed trolley runs on rails between them.

FAN-CONTROLLED FLOOR-FED PIGGERY

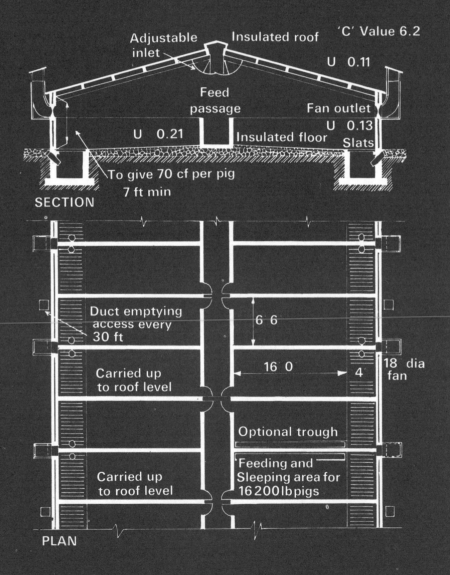

'C' Value 6.2

Adjustable inlet

Insulated roof

U 0.11

Feed passage

Fan outlet

U 0.21

Insulated floor

U 0.13

Slats

To give 70 cf per pig

7 ft min

SECTION

Duct emptying access every 30 ft

6 6

Carried up to roof level

16 0

4

18 dia fan

Optional trough

Carried up to roof level

Feeding and Sleeping area for 16 200 lb pigs

PLAN

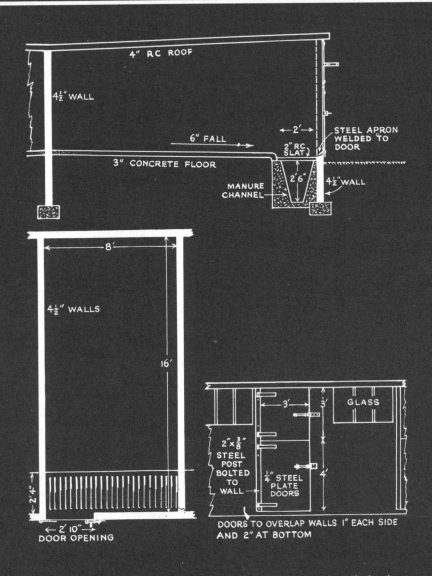

4" RC ROOF

4½" WALL

6" FALL

3" CONCRETE FLOOR

← 2' →

STEEL APRON WELDED TO DOOR

2" RC SLAT

MANURE CHANNEL

2' 6"

4½" WALL

8'

4½" WALLS

16'

2' 4"

← 2' 10' →
DOOR OPENING

3'

3'

GLASS

2" × ⅜" STEEL POST BOLTED TO WALL

¼" STEEL PLATE DOORS

4'

DOORS TO OVERLAP WALLS 1" EACH SIDE AND 2" AT BOTTOM

Slatted-floor piggery erected by Mr J. Jordan of Northern Ireland

DEEP-DUNG
FATTENING HOUSE

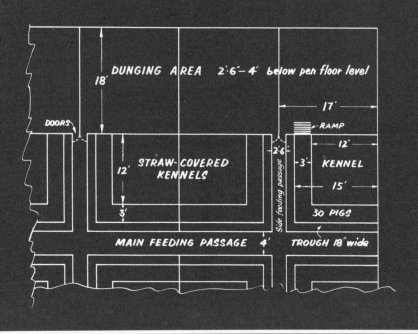

DUNGING AREA   2'6"–4' below pen floor level

18'

DOORS

12'

STRAW-COVERED
KENNELS

3'

2'6"

Side feeding passage

17'

RAMP

12'

3'

KENNEL

15'

30 PIGS

MAIN FEEDING PASSAGE   4'

TROUGH 18" wide

FACE-TO-FACE
HORVAT-TYPE PIGGERY

10' SHEET

5"x 2" PURLIN

3" KERB
BOARD

STY FLOOR

3" STEP

4" FALL

BRICK PIERS,
TIMBER OR
STEEL ANGLE
SUPPORTS

SLATTED FLOOR
FATTENING HOUSE

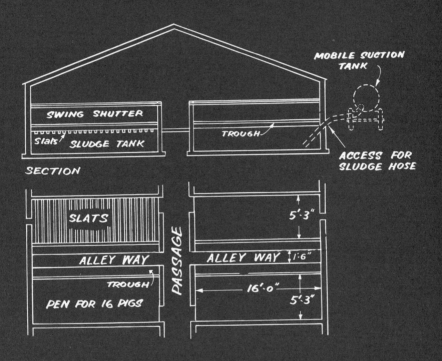

This type of house would need to be fully insulated and would house 64 pigs per 15' frontage.

SOLARI-TYPE
FATTENING HOUSE

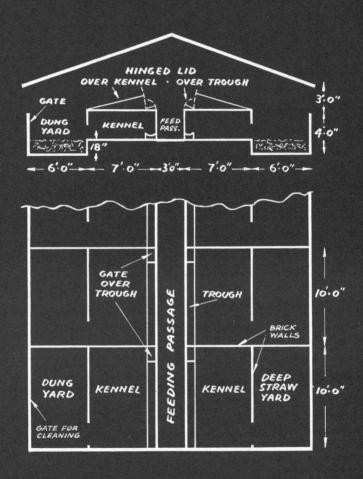

GARCEAU "SQUATTIE"
YARDED FATTENING HOUSE

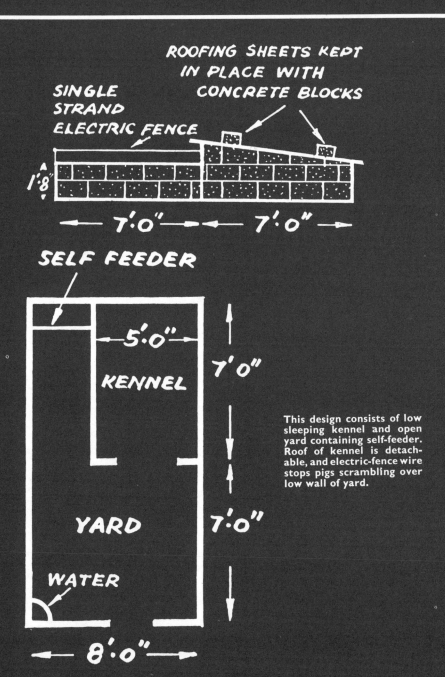

ROOFING SHEETS KEPT
IN PLACE WITH
CONCRETE BLOCKS

SINGLE
STRAND
ELECTRIC FENCE

1'8"

← 7'0" → ← 7'0" →

SELF FEEDER

5'0"

KENNEL

7'0"

YARD

7'0"

WATER

← 8'0" →

This design consists of low sleeping kennel and open yard containing self-feeder. Roof of kennel is detachable, and electric-fence wire stops pigs scrambling over low wall of yard.

OPEN-FRONTED
FATTENING HOUSE

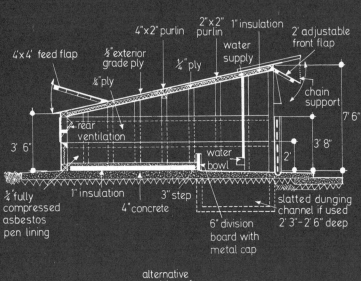

4'x4' feed flap

½" exterior grade ply

¼" ply

4"x2" purlin

2"x2" purlin

1" insulation

water supply

¼" ply

2' adjustable front flap

chain support

7' 6"

rear ventilation

3' 8"

3' 6"

water bowl

2'

¼" fully compressed asbestos pen lining

1" insulation

4" concrete

3" step

6" division board with metal cap

slatted dunging channel if used 2' 3"–2' 6" deep

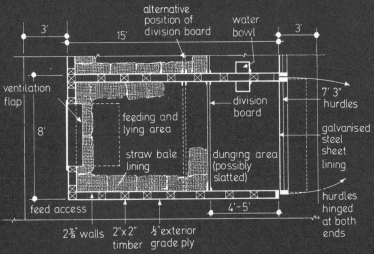

3'

15'

alternative position of division board

water bowl

3'

ventilation flap

7' 3" hurdles

8'

division board

galvanised steel sheet lining

feeding and lying area

straw bale lining

dunging area (possibly slatted)

hurdles hinged at both ends

feed access

4'–5'

2⅝" walls

2"x2" timber

½" exterior grade ply

YARDED PIGGERY

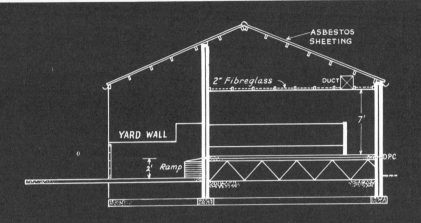

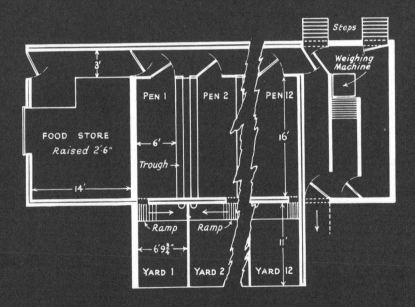

This design has a deep-dung covered yard, and each pen holds 16 fatteners.

FATTENING HOUSE WITH SIDE TROUGHS AND SLATTED DUNGING AREA

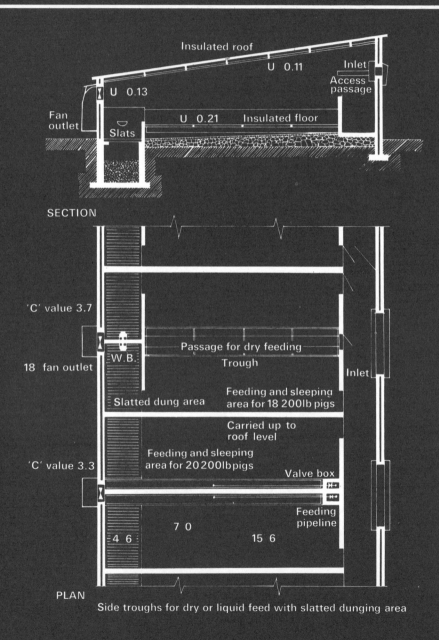

SECTION

Insulated roof
U 0.11
Inlet
Access passage
U 0.13
Fan outlet
Slats
U 0.21 Insulated floor

PLAN

'C' value 3.7

18 fan outlet

W.B.

Passage for dry feeding
Trough

Inlet

Feeding and sleeping area for 18 200lb pigs

Slatted dung area

Carried up to roof level

Feeding and sleeping area for 20 200lb pigs

'C' value 3.3

Valve box

Feeding pipeline

7 0

4 6

15 6

Side troughs for dry or liquid feed with slatted dunging area

TYPES OF
SLATTED FLOOR

T-BAR SLATS

Slats $1'' \times 1'' \times \frac{1}{8}''$. Gaps $\frac{3}{8}''$ and $\frac{3}{4}''$ alternatively. Supported by: $2'' \times \frac{3}{8}''$ flat.

REINFORCED CONCRETE SLATS

Slats $3'' \times 3''$. Gaps $\frac{7}{8}''$.

WELDMESH

Weldmesh $6'' \times \frac{3}{4}''$ mesh $\times$ 5 G. supported at $18''$ centres (for pigs three months and over).

Weldmesh $3'' \times \frac{1}{2}'' \times$ 10 G. supported at $12''$ centres (for weaners and older).

PRE-CAST CONCRETE SLATS

Reproduced by kind permission of Farm Buildings Dept., North of Scotland College of Agriculture.

SOW YARD WITH INDIVIDUAL FEEDERS

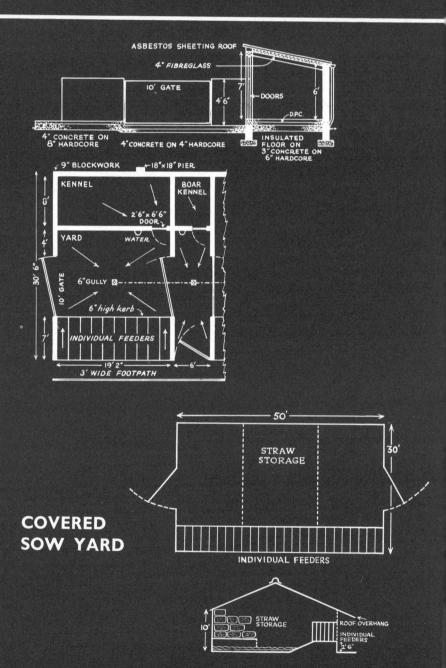

ASBESTOS SHEETING ROOF

4" FIBREGLASS

10' GATE

7'

4'6"

6'

DOORS

D.P.C.

4" CONCRETE ON 8" HARDCORE

4" CONCRETE ON 4" HARDCORE

INSULATED FLOOR ON 3" CONCRETE ON 6" HARDCORE

9" BLOCKWORK

18"×18" PIER

KENNEL

BOAR KENNEL

8'

2'6"×6'6" DOOR

YARD

WATER

4'

30'6"

10' GATE

6" GULLY

6" high kerb

7'

INDIVIDUAL FEEDERS

19'2"

6'

3' WIDE FOOTPATH

50'

STRAW STORAGE

30'

INDIVIDUAL FEEDERS

COVERED SOW YARD

STRAW STORAGE

ROOF OVERHANG

10'

INDIVIDUAL FEEDERS

2'6"

STALL FOR
TETHERED SOW

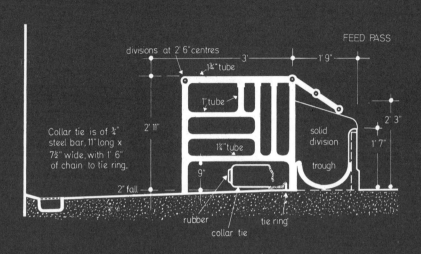

FEED PASS

divisions at 2' 6" centres

1¼" tube

3'

1' 9"

1" tube

2' 11"

1¼" tube

solid division

2' 3"

1' 7"

trough

Collar tie is of ¾" steel bar, 11" long x 7½" wide, with 1' 6" of chain to tie ring.

2" fall

9"

rubber

collar tie

tie ring

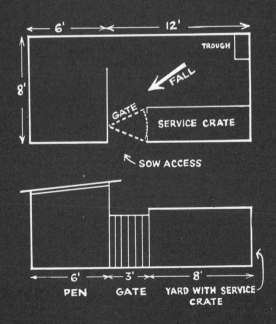

6'

12'

TROUGH

8'

FALL

GATE

SERVICE CRATE

BOAR PEN
AND YARD

with Built-in
Service Crate

SOW ACCESS

6'

3'

8'

PEN

GATE

YARD WITH SERVICE CRATE

STRAW-BALE HUT

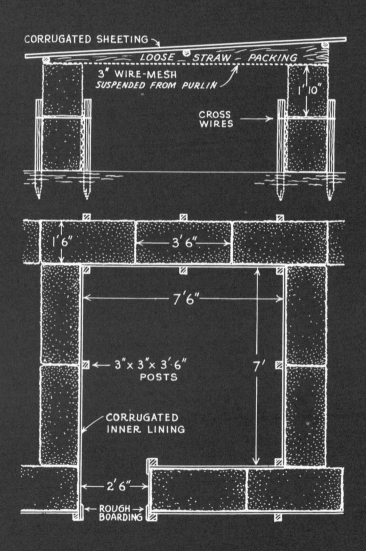

CORRUGATED SHEETING

LOOSE STRAW PACKING

3" WIRE-MESH
SUSPENDED FROM PURLIN

1' 10"

CROSS
WIRES

1' 6"

3' 6"

7' 6"

7'

3" x 3" x 3' 6"
POSTS

CORRUGATED
INNER LINING

2' 6"

ROUGH
BOARDING

FITTINGS

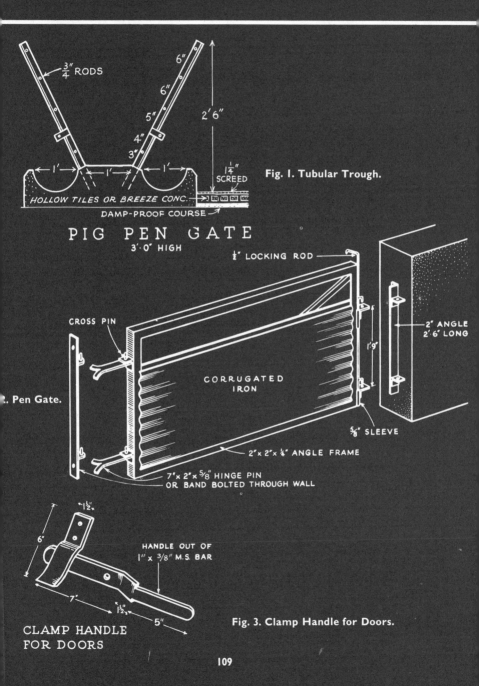

$\frac{3}{4}''$ RODS

6"
6"
5"
4"
3"

2' 6"

$1\frac{1}{4}''$ SCREED

1'   1'   1'

HOLLOW TILES OR BREEZE CONC.

DAMP-PROOF COURSE

Fig. I. Tubular Trough.

PIG PEN GATE
3'·0" HIGH

$\frac{1}{2}''$ LOCKING ROD

CROSS PIN

2" ANGLE
2' 6" LONG

I' 9"

CORRUGATED
IRON

$\frac{5}{8}''$ SLEEVE

. Pen Gate.

2" x 2" x $\frac{1}{4}''$ ANGLE FRAME

7" x 2" x $\frac{5}{8}''$ HINGE PIN
OR BAND BOLTED THROUGH WALL

$1\frac{1}{2}''$

6"

HANDLE OUT OF
1" x 3/8" M.S. BAR

7"

$1\frac{1}{2}''$

5"

CLAMP HANDLE
FOR DOORS

Fig. 3. Clamp Handle for Doors.

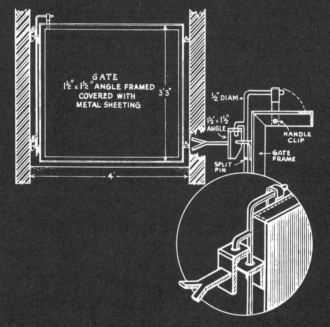

GATE
1½" × 1½" ANGLE FRAMED
COVERED WITH
METAL SHEETING

3'3"

4'

½" DIAM.

1½" × 1½"
ANGLE

SPLIT
PIN

HANDLE
CLIP

GATE
FRAME

Fig. 4. Pen Gate.

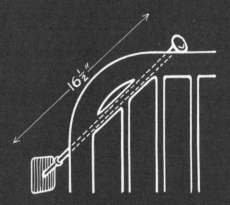

16½"

Fig. 5. Pen Gate Fastener.

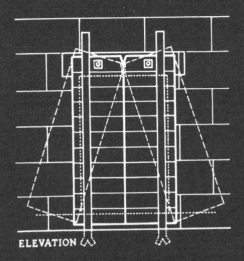

Fig. 6. Self-closing Doors.

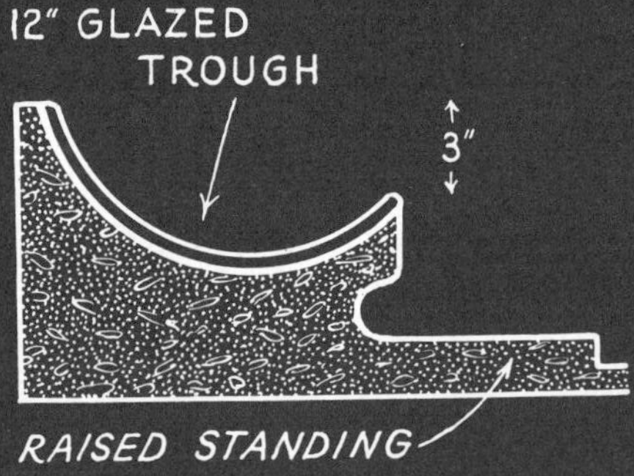

ELEVATION

Fig. 7. Glazed Trough.

12" GLAZED TROUGH

3"

RAISED STANDING

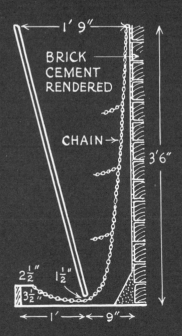

1' 9"

BRICK CEMENT RENDERED

CHAIN

3' 6"

$2\frac{1}{2}$" $1\frac{1}{2}$"

$3\frac{1}{2}$"

1' 9"

Fig. 8. Self-feeder.

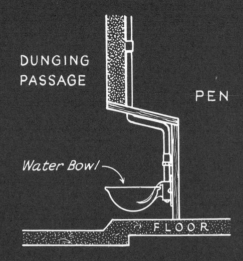

DUNGING
PASSAGE

PEN

Water Bowl

FLOOR

Fig. 9. Siting of Water Bowl.

1½" DIAM. TUBING

14"
TROUGH

Fig. 10. Swinging Front.

or barn which must be a warm structure. Each pen may be surrounded by a dwarf wall to retain the piglets. By rotating the crates through different buildings the risk of disease is greatly minimised, and a truly adaptable arrangement is available.

One of the disadvantages of the crate is that as the sow lies down she may make rather a sudden and heavy fall between the middle rail and the floor, since the bottom rail is recessed towards the wall. To avoid this, Mr Bunker has devised a hinged metal flap resting on the bottom rail which the sow lies against as she goes down but which hinges upwards as she rises. The detailed design is shown in the lower diagrams (page 82).

Another method of achieving the same end, and also coping with the considerable variation in size between different sows, and especially between sows and gilts, is to make the bottom rail in the crate variable in position so it can be moved a few inches vertically or horizontally. Several proprietary crates incorporate this feature with advantage.

SLATTED-FLOOR FARROWING PENS

There has recently been a considerable interest in farrowing pens with either part slatted or entirely slatted floors. In the case of the latter a suitable slat has been either a $3''$ or $4''$ concrete one with a normal gap of $\frac{1}{2}''-\frac{2}{3}''$ but with an enlarged gap of $\frac{4}{5}''-1\frac{1}{8}''$ in a 2 sq.ft. area behind the sow. For the first week after farrowing the area behind the sow is covered with expanded metal to prevent the baby pigs catching their feet in the gap.

Two forms of part-slatted pens have been tried successfully. In one case $3''$ slats have been used with $\frac{1}{2}''$ gaps, and the rear half of the pens are slatted. In another case a similar area is used but expanded metal is used—$1\frac{3}{4}''$ longways and $\frac{1}{2}''$ across with the grid supported every $1'$ by metal cross-pieces. It is best to use galvanised metal.

Whilst it is early days to state whether the fully slatted pen is likely to be a success, the results so far are promising. Those who have used this system have established the importance of keeping the piglets, and indeed the sow, warm since there is no

H

bedding—or at the most a modest amount for the piglets, in the early stages. It is also especially important that the slats, if concrete is used, should be impeccably finished, otherwise both the sows and the piglets will suffer.

RUAKURA FARROWING RING

The straightforward crate of parallel bars or sides is certainly the simplest arrangement, but other more complicated designs exist. A design which has achieved some popularity is the New Zealand or Ruakura farrowing ring. This is a farrowing pen of circular shape with a central creep. Because of the curvature of the pen, the sow can in theory only lie with her back to the outer curve. Thus her belly and teats will face the centre and the creep where the piglets will lie in warmth and safety. Though this design has been used in totally-enclosed buildings, it was really conceived for use as a voluntary crate-type arrangement in the shape of an outside hut. It rather misses the point used inside as the sow will have to be confined to it for a period or she may farrow outside the unit, and in area the 'Ruakura' ring and the exercise area would occupy more or less the same as required for two farrowing crates.

However, a number of adaptations of the design have been used and are marketed commercially. These are useful arrangements and are generally similar to the type shown on page 84. By 'squaring off' the design a more economical shape is achieved though the arrangement has the same advantages as the original circular ring. Doors are incorporated between the sow's lying area and the exercising and feeding area (as it is still found necessary to confine the sows during farrowing time). When the sow is let out to exercise the piglets are confined to the creep and sow's lying area by dropping in a 10″ high board across the bottom of the doorway which only the sow can step over. Naturally it is good to allow the sow as much freedom as possible and as soon as she can be allowed to go in and out of the lying area as she pleases, so much the better.

A delightfully simple farrowing arrangement is the Challow timber farrowing nest which can be placed in an existing pen. It measures 7' 4" × 4' 2" and is only 1' 2" high. An open area is provided in the centre in which the sow is free and there are warm and safe side areas covered with hinged flaps for the piglets.

The Combined Farrowing and Rearing Pen

When a specialised crate is used, such as we have described in the previous section, the sow and litter stay in it for only about ten days to a fortnight, no provision being made for giving creep feed. This means that the piglets and sow must be moved on to a rearing pen, again of specialised design, providing a creep for the feeding and warmth of the piglets. Some may rightly question the desirability of a move at this period which may cause stress and check the growth of the piglets and the milking ability of the sow. The crate also forms a bottleneck in the cycle and also a particularly dangerous potential source of disease build-up.

In the larger herd it is not difficult to provide for a range of crates and rearing pens in the ratio of 1:2 where the sow and piglets will spend a third of the time in the crate and two-thirds in the rearing pen. In the smaller herd, however, it is too expensive to provide these facilities; in these cases, and indeed in any herd where both flexibility and an absence of moving the sow and litter is demanded, the combined farrowing and rearing pen is a great advantage (see page 83).

The commonest arrangement is a pen having a 10' frontage and 8' in depth. Down one side of the pen is the combined creep and nest, which has a depth of 2' 6". Next to this is the crate which is formed as a temporary arrangement by having a gate of boards or rails 2' 6" from the creep rails and parallel to it. The second row of rails may be pivoted at the top so that it can be pulled up and over the creep after the first ten days to a fortnight; or the rails may be made in two sections in the form of a pair of gates and fixed at the centre. The latter arrangement provides gates that are easy to open and close when the sow is being exercised, but obstruct the pen rather more when not in use.

The only other fittings in the pen are the sow's trough and

water-bowl in the front. The pen frontage is therefore taken up by 2′ 6″ of creep, 2′ 6″ of crate, 2′ 6″ for the trough and water-bowl, and 2′ 6″ for the doorway. Access to the creep is satisfactory but in order to facilitate it, it may be preferred to have a small catwalk of 1′ 6″ between alternate pens.

SIMPLE, HYGIENIC LAYOUT

The simplicity of this design will be apparent. Also, it can form a desirably hygienic layout. There is no contact between pens as the drainage can run outside in a small open channel at the side of the passageway. A good fall on the floor will take the urine and water to the corner of the doorway and out to the channel. The open channel may seem to some to be an undesirable feature, but in reality it is probably a much more hygienic arrangement than a closed drain running under the passage. Drainage to the back of the pen and a trap to the outside will give good hygienic drainage, but this is an expensive arrangement.

A satisfactory layout with this type of pen is a double row of pens with a centre 4′ wide passageway, the building being approximately 20′ wide. A unit of up to eight of these pens is the maximum that one would place in one unit without a complete cross-partition of the building, providing the means of closing it down and fumigating it frequently.

The Solari Farrowing Pen

A simple approach to farrowing and rearing is epitomised in the design that aims to keep litters warm and comfortable by providing a limited air space and freedom from draughts, and keeps the sows in check by judicious use of farrowing rails and a limited area for the pens.

The pens consist of a range of units each 5′ wide and 16′ 6″ long. The height of the back of the pen is just 3′ rising to 8′ at the front. In the front there is only a wall and a door, both 4′ high, leaving a 4′ high open area above to allow free air circulation. There is a creep of 2′ 6″ depth the whole way across the back, formed by a simple steel grille. In front of the creep, farrowing rails are fitted extending 5' 10" in front of the creep; there is a space of 2' 9" between them and the sow lies voluntarily in this area. This leaves the rest of the pen for exercising, feeding and dunging. A photograph is shown between pages 72 and 73.

The rear part of the pen floor only is insulated to encourage the sow to lie in this area, and there is a good fall from back to front of 5". Other important dimensions are that the farrowing rails at the front and back of the 'crate', if it may be called that, are 6" from the floor, whilst the side ones are 10". The whole arrangement is easily demountable; when using the pen for farrowing all the equipment will be in place; after ten days the rails will be removed leaving only the creep grille. Should it be desired to continue using the pen thereafter the creep too may be removed.

ADAPTATION FOR PORKERS

One clever adaptation of this pen for use from birth to finishing is to place a slatted or weldmesh area in the front which is covered during the farrowing and rearing stage but uncovered afterwards when the sow is removed and no litter may be used. A pen could take up to 20 porkers, so the cost of this accommodation would be extremely reasonable.

The system is altogether simple and straightforward. Its disadvantages are that environmental controls are incomplete and labour requirements are quite heavy. The pigman also has a less favourable deal than in the conventional totally-enclosed house; he has no protection for tending the pigs and access to the nest and creep are behind the sow and therefore more difficult.

Farmers who have erected this type of house in exposed areas have found the pens can become too cold in the winter and the water-bowls freeze. Under these circumstances there is obviously a temptation to have a totally-enclosed unit merely by adding a well-ventilated covered service passage along the front of the pens. It also clearly does not give the complete control over the sow that the crate does.

117

'Isolation Housing'

Many farmers have felt that one of the drawbacks to the conventional housing of pigs is that it involves too many moves, creating a check or a stress on each occasion that can retard growth for a period and produce such a stress that it may lead to the onset of disease. There is undoubtedly much truth in this as outbreaks of disease, such as enteritis and bowel oedema, will frequently follow changes in environment and housing. On top of this I have already emphasised the fighting and bullying that takes place when pigs are mixed except in the earliest stages; mixing pigs may upset their 'bacterial equilibrium' and is reason enough in itself for limiting movement.

In earlier chapters we have shown the feasibility of economically reducing the moves in a fattener's life to a single one—from rearing pen to fattening pen. Several breeders have, however, produced a pen which can be used the whole lifetime of the pig. Their observation is that a litter reared intact, never mixed with other pigs, will always do best.

The pen as a unit in this type of building is usually separated from its fellows on either side by a complete wall from floor to ceiling isolating not only the pen but the dunging area; this does make for one real disadvantage, namely, that dunging out is usually by hand and virtually a shovel-and-barrow job. The pen itself to be sufficient for a sow and ten pigs needs an area of approximately 10' × 8' with 10' frontage giving trough space for 11 baconers. The yard would be 10' × 8'. In a pen this size there is room for one of the crate designs of a temporary nature and a creep along the front of the pen that will allow young growing pigs access but not the sow. One scheme is to feed the sow separately in the yard if she stays with her progeny right through their fattening period and her next pregnancy.

If access to this type of housing was from an outside but covered way with doors to each pen, it would give us true isolation units but at considerable cost. The more usual procedure is to provide a common passageway as part of the building but to be divided by a few doors along its length, so that whilst there is not perfect isolation, pigs are neither mixed nor moved from birth to finishing. Others run the system rather differently, merely keeping the litter together as a unit and taking the sow away after weaning to be managed in pregnancy in the usual way.

Since the first edition of *Pig Housing* was published there has been considerable interest shown in 'isolation', or as it is more commonly known now, 'farrow-to-finish' housing. The Agricultural Research Council conducted an investigation* comparing litters reared in 'farrow-to-finish' pens with those subjected to the usual move from farrowing and rearing pen to fattening pen together with the mixing of two litters together. It was found in this extensive trial that 23 days were saved in the period from birth to bacon weight by keeping the litter together in one pen. The type of pen used in this trial measured 9' 8" wide and 8' 6" deep with the outside 2' 6" of floor area slatted. A temporary crate and creep area was used during the farrowing and rearing period and the slats were covered with plywood during the first few days of the piglet's life. The fatteners were floor-fed.

This design (Fig. 20), or alternatively with a solid-floored dung passage, forms the most popular farrow-to-finish shape, but pens basically of Solari and farrowing pen shape are also popular and may be placed in the usual lean-to housing or under a covered yard with straw storage above.

Farrowing Outdoors

The easiest farrowing accommodation to manage is certainly the indoor all-enclosed unit, but the merits of outdoor farrowing are not inconsiderable. One of the problems that many farmers have experienced in their all-enclosed quarters has been chronic disease; this almost never occurs except sporadically when sows are farrowed outdoors. Outdoor farrowing units are also cheaper in initial cost even if their life is comparatively short.

*The Housing of Pigs in One Pen from Birth to Slaughter, Charlick, R. H., Livingstone, H. R., McNair, A., Sainsbury, D. W. B. Experimental Farm Buildings Report No. 11, N.I.A.E., Wrest Park, Silsoe, Bedford.

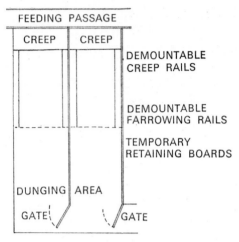

Fig 20. *Farrow-to-finish pen design used at the School of Veterinary Medicine, Cambridge.*

A hard-wearing unit for outdoor farrowing is the ark form. The Craibstone ark (designed by the Farm Buildings Department of the North of Scotland College of Agriculture) is an excellent unit of this sort. Dimensions are 8' long, 7' wide and 5' high (see page 90). The shape of the hut itself to some extent forms a safe escape for the piglets where the roof meets the floor. The shape also gives it rigidity and cheapness. Some of the essential features of outdoor farrowing huts, of which several types exist, are well shown in this hut but should be incorporated in any farrowing unit.

CRAIBSTONE ARK CONSTRUCTION

The first essential is sound, two-skin construction, which ensures warmth and freedom from draughts. Next, really good baffling is needed for the pigs' doorway. In the Craibstone ark a baffle is placed inside the door to deflect the air away from any opportunity of causing draught on the nest. In other designs a porch is made on the outside with double sacking curtains so that something approaching an air-lock is reached between inside and outside environments. The huts must also incorporate a nest and creep-feeding and watering arrangements when the building is to be used for rearing. To keep the piglets warm in

One of a range of multi-suckling pens fitted with movable divisions to extend pen size as pigs grow.

In this specialised weaner house [two or three litters are grouped together in box-like kennels with an outdoor wire-mesh run.

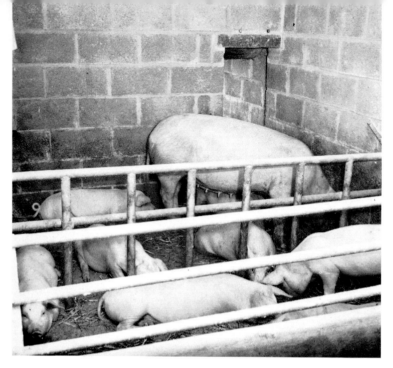

Sow with her litter in a birth-to-slaughter pen. Tubular barrier prevents her reaching the pigs' trough.

Wet-feeding fatteners by pipeline on a Dorset farm. Quarter-turn outlet valve is operated by a key.

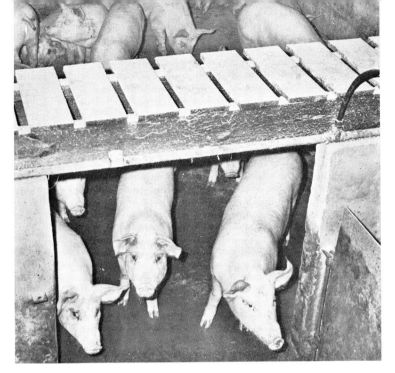

Catwalk running along pen tops on one side of a central dunging passage replaces the feeding passage.

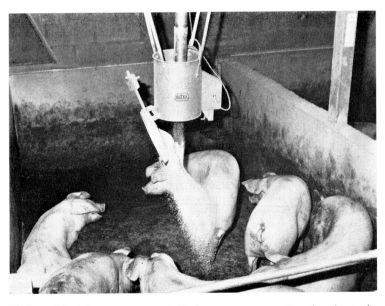

Mechanical feed dispenser automatically drops a pre-set quantity of meal onto the pen floor.

Controlled-environment fattening house for floor-feeding. Air is drawn in through the ridge and extracted at floor level through the pop-holes.

Below: Pen-fronts in this house are adjustable, limiting floor space to encourage clean habits.

their nest and also to encourage them into it, straw is placed on top of it. Artificial heating is difficult in the field but is commonly used when huts are sited on permanent standings.

One of the general criticisms of outdoor farrowing and rearing is the difficulty of giving proper attention to the sow and the litter in the event of disease or mishap. To help answer this criticism a good door for the pigman should be provided at the back 2' wide at least and extending 3' 6" high. Though this will be a great help, the fittings on the door must be extremely tight-fitting to give a complete seal to the door when closed (see chapter 13). In the Craibstone ark, and others, there may also be a small observation door up to 1' square in the back or elsewhere that can be used simply for quiet inspection without disturbing the inmates. It is usual to fit a wooden batten to the floor 2" wide and 1½" deep along the sides 9" from the edge, to prevent the sow's feet jamming the piglets against the wall and causing them injury.

The idea of the Ruakura ring has also been very successfully incorporated in huts of proprietary manufacture with central-placed creep. This indeed is how the ring was originally intended to be used and has been used with great success by a number of farmers. Round huts of metal or plywood construction are robust, hygienic and warm, also light and reasonably easy to move.

SLOPING FARROWING FRAME

A design of originality is also seen in the Harper Adams 'Sloping Farrowing Frame' which, whilst it can be incorporated in a hut, can also be used as an alternative to the conventional crate. As devised by Mr W. T. Price, the sloping farrowing frame is a wooden structure, essentially a crate with wooden sides sloping inwards at an angle of approximately 25°, the bottoms of the boards being 10" from the ground, the top 3' 6" and the width at the bottom being 3' and the top 5' 6". A unit like this can be placed quite conveniently in a rectangular farrowing hut.

The sloping frame has the same advantage as the Ruakura

ring in that the sow is not confined or shut in—her movements in and out being voluntary. Piglets may, however, always be confined to the inside for the first few days by placing a 1' high board across the bottom of the doorway. The advantage of the sloping frame is that the sow will move downwards gradually and will not be inclined to the rather sudden drop that takes place in a farrowing crate between the middle rail and the floor and which may crush a piglet.

RECTANGULAR HUT WITH CREEP

Another common type of farrowing hut is of rectangular shape and measuring 8' long and 5' 6" wide plus a further 2' extension at the back to form the special creep. Height at the front is 4' and at the back 3'. The general principles I have mentioned in connection with the ark will also follow, but as well as baffle doorways, creeps and nests there must also be a farrowing rail 10" from the floor and 10" from the wall. With both farrowing huts and arks there may be a fold unit in front, usually about 12' long, or the sows may run in paddocks or be tethered. In normal ways the use of a specialised farrowing hut of one of the forms described here, or adaptations of it, is advisable.

ROADNIGHT SHELTER

For pure simplicity, the arrangement followed by Mr Richard Roadnight has no peer. He uses simple curved corrugated sheets as the only protection for the sow and piglets, with one end blocked up. There is copious bedding but no floor. Sows are put into a field in groups to farrow together and choose their own accommodation. A system like this costs virtually nothing in housing but it needs a large acreage providing not more than three sows to the acre, good drainage, and preferably Saddleback sows. Also the farrowing can only be done in the warmer months—say, from March to October as outside limits. It is one of those systems that gives excellent results in the originator's hand and will do in others who understand the principles and integrate it into a general farming routine under the right circumstances. The copier who understands the system only in part will generally fail.

OUTDOOR PIG MANAGEMENT

It is interesting to record that at the time of the production of this, the third edition of *Pig Housing*, in 1972, there is a renewed interest in outdoor pig management in suitable climatic areas with light and well drained soil. That it can be profitable has been shown by the work of Boddington at Wye College.* There are now several successful units of up to 250 sows in parts of Norfolk, Kent and Berkshire which employ simple housing, such as we have described, and the signs are that this trend, which to some extent goes against the main pattern of development of indoor housing, will continue as a useful, profitable and healthy alternative.

---

* Outdoor Pigs: Results of an Economic Investigaton by M. A. BODDINGTON. SCHOOL OF RURAL ECONOMIC AND RELATED STUDIES, Wye College, Kent.

Chapter 8

REARING PENS

THE fullest protection for the piglet is required for the first ten days to two weeks of life, when both crushing and chilling are at their commonest. The crate or similar arrangement is almost essential equipment. Thereafter, if the convertible type of pen described in the last chapter is not used, the piglets must be moved to good accommodation with warmed creep. A suitable layout is shown on page 85. In a house with a central feeding and service passage, pens are placed on either side, of dimensions 10' × 8', the longer side running along the front. The creep is placed along the front of the pen, measuring 5' × 3' minimum, and adjoining it is the sow's trough and water-bowl and a 2' wide gate. Only a dwarf wall 2' high is needed adjoining the creep—enough to prevent draught but allowing ease of access.

In some cases there is a dung passage behind the pen similar to the Scandinavian design of fattening house, but it is questionable whether this is a good feature. For one thing, there is a danger in a common dung passage, of moving infections between pens with greater facility when the passage is cleaned out. For another, there is not a great deal of muck to clean out in a rearing pen and the incorporation of a dung passage very materially increases the cost of a building, its cubic area and the difficulty of heating it.

Using a central service passage, drainage can be by open

gulleys on either side of this so that there is no contact between pens. To ease the problem of cleaning some passages have been made wide enough to enable a tractor and trailer to be taken through. This is a procedure to be deprecated as the building is made much bigger, environmental control more difficult, and the disturbance created by the tractor and trailer is a stress to be avoided. However, one way of easing the cleaning problem, without introducing disturbance, is to have a form of automatic drain either in the front of the pen with slatted floor, or under the passage, so that the muck can be cleaned out in semi-liquid form (see description of drainage in chapter 9). This does, however, entail having virtually a litter-less pen, and many will not accept this as good husbandry. An alternative is to have a scraper-cleaned gulley outside the pens which will take muck and litter in solid form; this has been applied successfully on a number of farms. The fall in the floor towards the end and corner taking the drainage away should be a good one, of the order of 4″ from back to front and 2″ from side to side.

VENTILATION ON SIMPLE LINES

The ventilation of this type of unit can be on simple lines with hopper type of window inlets and overhead extraction. A disadvantage is that the control is made on the windows at the back of the pen. A simpler arrangement, both of pen and ventilation, has been used by James Watson, formerly of Muir of Pert Farms, Dundee. His is a narrow building with side service passage and pens on one side only (see page 87). The pens are 13′ deep and 6′ 3″ wide, with the ceiling falling towards the back, where the creep and nest are situated. Otherwise the pen is similar in conception to the arrangement previously described, but air is extracted over the wide passage by chimney ventilators assisted by fans drawing in fresh air, or from louvred air inlets in the wall below; the pen being ventilated only by induced convection currents.

A further principle of this design is that of inter-pen isolation by continuing the dividing walls to the roof which in fact rests on them. Whilst the design is, therefore, simple and inexpensive, it has a disadvantage that the creep is placed at the back of the pen so that the stockman must walk through the pen and past the sow to get to it. Good stockmen, however, like to walk amongst their pigs and may consider this actually an advantage!

These two designs have been of totally-enclosed buildings, which are generally acknowledged to be most satisfactory, but for the farmer who wants an outside yard, believing in the merits of some fresh air and sunlight for sow and litter, there is a suitable design basically of similar form to the rearing pen on page 85, but a slightly smaller pen is suggested, as dunging and some exercising will be outside. The position of the creep is, however, of very great importance as it must be sited to prevent draughts blowing on to it. This is inevitably the great difficulty with all accommodation with outside yards and must be dealt with by complete inter-pen isolation, and good baffle walls, doors or curtains between the pen and the yard (see chapter 9).

The position of the creep in this design is chosen to run along the wall parallel to the entrance between pen and yard because here draughts are least likely with the solid end of the creep serving as the inside baffle of the doorway. The creep is, however, so placed that it is still easily cleaned and food and water can be added from the service passage, without entering the pen. In all cases where the creep does adjoin a service passageway or 'catwalk', the wall into the creep need be no more than 2' in height. The opposite wall should be 3' 6" high, preferably of solid construction with this design, and covered with a solid hinged top. The building should be kept as low as possible and a height of 7' at the service passage side is sufficient.

In many of the rearing pens it is popular to provide pop-holes connecting the creeps. After the litters have settled to their own pens, the pop-holes can then be opened to allow the piglets to mix, thus preparing them for later weaning and reducing the 'mixing stress' at this time. This is a satisfactory arrangement, particularly for the farmer who prefers to avoid mixing several sows as with the multiple suckling pen.

Multiple-suckling Pens

A further major change that has taken place in pig management since this book was first written, is the popularity of the 'multiple-suckling pen'. One of the major problems of pig husbandry which I have referred to earlier is the serious check that takes place when piglets are weaned and mixed at the same time. One approach is the 'farrow-to-finish' pen where pigs are taken from birth to finishing in the same pen. Another is to mix three to five sows and litters together, usually at about three weeks of age and thereby form a 'weaner pool' which includes the dams. After two or three weeks the sows are removed and the group of 30 to 50 weaners are left for a further period almost always *ad lib* fed until ready for the finishing stages of fattening.

There is no doubt that weaners mix with more ease if they are still receiving their mothers' milk, and their young age also seems advantageous. The sows also may benefit from the mixing at this stage. Considerable flexibility is possible with the housing which may be totally enclosed or yarded and proven examples of which are shown in figs. 20 and 21. the first being of an open-front yard form and the second in a totally-enclosed but deep-strawed building.

The essentials are as follows: adequate space, and it is recommended to have at least 60 sq ft for each sow and litter. Plenty of bedding is preferable under most systems and a large creep for the piglets of not less than 2 sq ft per piglet gives plenty of food and water space. The creep should also be warmed and covered in the usual way. Individual feeders for the sows are recommended wherever possible. Both the designs shown can be tractor-cleaned, the yards by access from the front after the feeders are removed and the totally-enclosed pen by access from the side after removing the pen divisions after each batch has gone through.

Early Weaning

These designs present one with a choice of arrangements for the 'rearing' or 'suckling' pens when the sow is present and

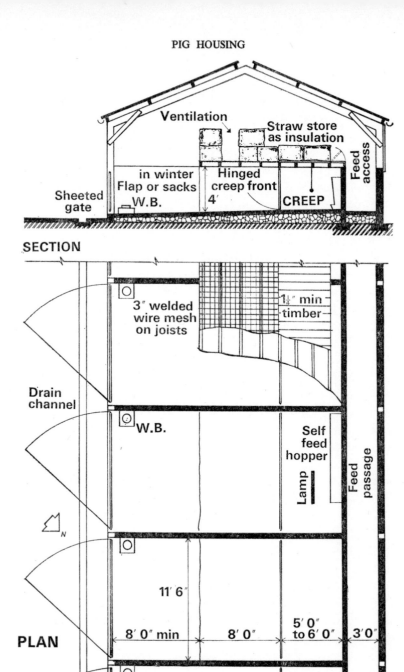

Ventilation

Straw store as insulation

Feed access

in winter
Flap or sacks
W.B.

Hinged creep front

4'

CREEP

Sheeted gate

SECTION

3" welded wire mesh on joists

$1\frac{1}{4}$" min timber

Drain channel

W.B.

Self feed hopper

Feed passage

Lamp

N

11' 6"

PLAN

8' 0" min

8' 0"

5' 0" to 6' 0"

3' 0"

Fig. 21. *Weaner pools and yards for 4 sows and litters (floor-fed sow cobs).*

128

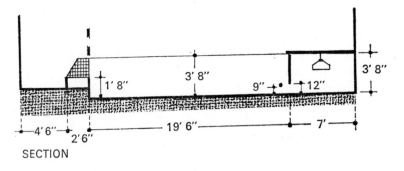

SECTION

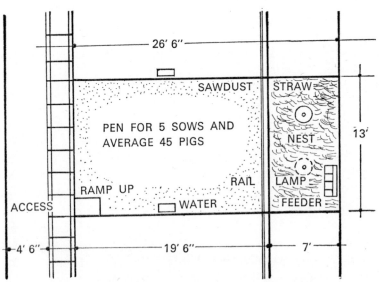

Fig. 22. Scottish design of rearing pen for five sows and litters.

which should offer a choice suitable for most enterprises, either as a new building or in the adaptation or conversion of an old one. The dangers of disease build-up are also as great in this age group as in the earlier farrowing pen, but even they are not as great as in the 'early weaning' pen which aims to house, as

129

efficiently and safely as possible, small pigs from 10 days onwards that have been weaned from their dam at this early age.
This system has had a chequered career, and it is in the disease
and housing fields that it has tended to fail, rather than on
nutritional grounds.

HAVE EASY-TO-CLEAN PENS

Piglets can be early-weaned quite successfully from three to
four days onwards with the adequate diets that are commercially
obtainable and provided they have received colostrum from the
dam for a day or two. The trouble has apparently always been
that when large numbers of piglets are taken through a building
at a very early age, with a diminished resistance to disease,
enteric disorders are very liable to occur, weights diminish and
mortality rises, so that the system becomes uneconomic. It is
essential, therefore, that buildings are made small enough to be
frequently emptied and cleaned and are hygienically constructed
so that disinfection and fumigation can be efficiently carried out.

A satisfactory design is shown on page 88. The pen measures
7' square. It is again desirable that it is built within a building,
keeping at least a degree of isolation between pens. Drainage
should be outside the pens in a gulley to the side of a central or
side service passage. A pen of this size would be quite sufficient
for a normal litter of 8–12. The pen incorporates a raised sleeping area, 4' square, which is at the back of the pen under the
lowest part of the roof, to conserve warmth and reduce draught.
It may be constructed of timber if easily dismantled, cleaned
and rested between batches. The pen also allows for reasonably
generous dunging, feeding and exercising area on two sides of
the raised sleeping area. Overhead artificial radiant heating is
required to a total capacity of not less than 50 watts per piglet
housed, even when the standard of insulation is high.

In order to avoid the problems of scouring and other infectious disease problems in young piglets, some farmers have
practised early weaning out-of-doors, in accommodation similar
to the 'hay-box' brooder used for chicks. A small wooden hut,
well-constructed and insulated, and in effect a rearing pen in

miniature, measuring 5' square with a 10' long run in front, is used.

STRAW-BALE HUT

Another popular method is the straw-bale hut, receiving one or more litters with an outside grass run. The hut should be considered a temporary structure which *must* be destroyed after a few litters have gone through. The separation of litters plus the generous space allowance that can be provided, and access to fresh air, all contribute to produce healthy results, but management can be difficult especially in winter and food requirements are likely to rise.

Rearing Pens with Verandahs

In recent years rearing pens consisting of kennels with outside slatted runs have become most popular. The simplest arrangement is simply a row of raised outdoor kennels opening onto slatted runs that allow the muck to fall onto a concrete pad from which it is removed from time to time.

Now, however, it is generally preferred to have two rows of the kennels separated by a feed passage and the whole of this section is covered. This makes the stockman's job much simpler. There are still slatted runs on each side of the building, serving the two rows of kennels and these are uncovered and usually placed over a slurry pit for periodic emptying. Pen size is usually 8' × 6', and the outdoor run is 8' × 4'. Capacity is about 25 piglets from 3 weeks at the earliest to 10–12 weeks at the latest. There is usually a drinker in the run and feeding may be on-the-floor or by *ad lib* hopper.

The system requires much more careful management than the strawed weaner house and the ventilation and environmental control are especially critical.

Cage Rearing

At the present moment the development of successful early weaning systems have presented new demands on housing piglets at their most critical stage. In the most popular system, the largest piglets in a litter are taken from the sow at seven days

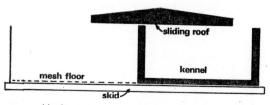

sliding roof

kennel

mesh floor

skid

Kennel fully insulated

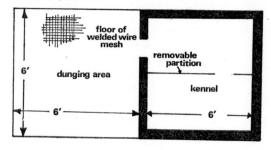

floor of welded wire mesh

removable partition

dunging area

kennel

6'

6'

6'

Fig. 23. *Hut-and-run weaner pen.*

wire mesh

slurry channel

Kennels fully insulated
Ventilation openings at eaves & ridge

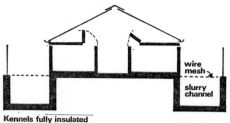

24'

4'

water

6'

3'

8'

dunging area (wire mesh floor)

kennel

passage

kennel

dunging area

Fig. 24. *Verandah-type rearing house.*

132

old, leaving their smaller litter-mates to catch up by taking the extra milk off the sow. Piglets are grouped by weight, nine at a time in cages three, four or five tiers high. Each cage measures 4' long $\times$ 2' wide $\times$ 1' 3$\frac{3}{4}$" high and has a floor of $\frac{1}{2}$" $\times$ $\frac{1}{2}$" $\times$ 12 gauge wire-mesh. These cage batteries are placed in housing kept at 80°F., and in subdued light. The piglets remain here until about 15 lb weight after which they are moved to flat-deck cages, each one being 4' $\times$ 4' and 2' high. The same floor mesh is used as in the first stage and piglets can be taken on here until about 60 lb weight.

Chapter 9

FATTENING ACCOMMODATION

The Totally-enclosed House

THERE is little doubt that the whole trend in pig housing is towards the totally-enclosed piggery where the environment is under complete control and all attendance to the pig is under cover. A basis of design and the most familiar is the Danish-style piggery with central feeding passage and side dunging passages (see page 91). In its conventional form it involves a central feeding passage 4' wide, side dunging passages 3' 6" and a pen of 10' × 6', to hold ten pigs to bacon weight. Alternative pen sizes to cope with different arrangements are described on page 144, and as is explained there, the conventional arrangement is probably the most expensive, in layout and cubic area per pig, of all designs. It remains an adaptable design, and is particularly suitable for the smaller pig-keeper. It has the advantage that pigs are kept in small groups, and management can be of a high standard. Feeding is simple and can be easily carried out with a trolley or overhead conveyor; dung cleaning may be by hand, squeegee, or by mechanical means. Straw may or may not be used, as desired.

In the original design the Danish piggery presented a large open air-space in which the dunging passage, pen and feeding passage were separated only by a 3' 6" high wall. Current practice is to screen off the dunging passage from the pen, leaving only a pophole between the pen and the passage. In this case the environmental control is very much improved, as the pigs in reality lie in a building within a building. In such a design, making use of good insulation and mechanical ventilation taking in fresh air from the ridge, temperatures within the range of 60° to 70°F may be obtained without difficulty. Only the central part of the house needs complete insulation. It is always best to place a number of cross-partitions across a Danish-type piggery and perhaps aim to have not more than 100 pigs within a common air-space and 300 to 400 pigs in one building.

General points on insulation and ventilation have been dealt with in the sections on these subjects. A popular practice is still to have a length of pen, from 10' to 15', taking 10 to 15 pigs, and a depth behind the trough of 5' or 5' 6". The pen itself should be raised 2" to 4" above the dung passage. The height to eaves need be no more than 6' to 6' 6" and a low pitch on the roof helps to conserve heat. Weighing is carried out either by weighing in the dung passage with portable scales, or by circulating the pigs round the building to a central or end weighing room. Whilst the former arrangement requires less movement of the pigs, the latter system is to be preferred because it enables a better and more accurate weighing-machine to be used; this can be placed in a well, the pigs going through it at ground level and the men weighing them are able to do the job and record, the weights more accurately (see page 182).

Wide-span Totally-enclosed House

In an attempt to produce a more economical piggery, though basically of the same layout and where trough feeding is required, a wider span building may be employed by running the troughs between the pens. In the Danish layout the cross-section is approximately 26', 7' being taken up by the dunging passages, 13' by the pens and trough and 4' by the feeding passage, making, with the walls, a total of 26'. If, however, we use the more economical span of 44', the dunging passages can each be 5', making 10' in all, the central service passage 4' and the pen depth may be 15' on each side, to take 15 pigs to bacon weight. The depth of the pen behind the trough will be 5' 6", the trough 1' and the catwalk serving the trough 1' (see page 93). The economy of this design is apparent when it is noted that whereas in each foot-run of a Danish house two pigs can be kept, in each foot-run of this type of house, four pigs are housed. The disadvantage is that the feeding system is slowed up as the pigman must move off at right angles to the main passage to put the food in the trough, but this problem is eliminated when an automatic feeding system is used either dry on-the-floor when no trough is needed, or with automatic pipeline arrangement using a trough. A popular modification is to make the dunging passages 7' wide to allow tractor cleaning.

In a design of this sort division of the building by cross-sections is particularly important to give the extra environmental control needed. A good arrangement is to place the divisions after every fourth pen, but as a minimum I would insist on a division after every eight pens, that is 120 pigs only in a section.

The construction of the pen division above 'pig height' can be of simple form; for example, a single skin of plywood, oil-tempered hardboard, or asbestos can be used satisfactorily. Doorways should also be installed along the central passageway; preferably double doors with two-way hinges and automatic closing, so that the trolley can be pushed through in either direction. With the 'reversed' ventilation arrangement of taking the fresh air into the centre of the house first and then passing it into and out of the dunging passage afterwards, there is no need for complete division of the dunging passages other than by the usual dividing gates to 3' 6" high.

CENTRE DUNGING PASSAGE

Many will rightly consider that the worst chore in the management of pigs—and indeed one that should be eliminated from the activities of a man as skilled as a good pigman—is dung cleaning. An easy way of helping towards this end is by the installation of one centre dunging passage, rather than a centre feeding passage, allowing access by the pigs from each pen to

Norwegian fattening house with complete slatted-floor pens and raised feed catwalks. The slats are of concrete.

Partially slatted-floor fattening pen.

Slatted dunging passages. Note sloping concrete fillets (photo, left) that keep the edges clean.

Fattening pen with dunging area of welded mesh.

Tubular metal barrier separating steel slats from lying area.

Example of the Suffolk-type feeding house based on a general-purpose building with pipeline trough-feeding and tractor scraping for manure.

Kennel-type fattening house, with covered feeding/dunging run.

half the length of the passage opposite their pen. Such an arrangement doubles the movement in feeding, which is usually twice a day, and halves the movement in dung cleaning, which is usually done once a day, so that in some respects the design has a disadvantage. This may not be considered so, however, by those who feel that the labour spent on feeding is time well spent and should not be so speedy that there is inadequate time to inspect the pigs, note their behaviour and health, and thus put production in jeopardy.

Ventilation can be simple and on traditional lines by extraction of stale air over the dunging passage. The dung passage may be screened off but there is less reason to do this than in the traditional Danish house.

It is noteworthy that the pigs have no contact at all with the outside walls of the house; this is quite an important point as the floors may well be warmer—much heat being lost laterally from the pig when it lies against the outside wall of a building. Nowadays, too, it has some real advantages as many houses will be of prefabricated design and lighter and less robust walling can be used though it will, of course, be as well insulated and vapour-sealed as in any other design.

There is yet another reason for adopting the centre dunging passage arrangement, particularly where slats are used and there is an automatic drain underneath. With a centre passage there is considerable saving in cost; instead of two runs of dung passage and slats there is only one, and the cost of this is little more than half as the width need be increased only slightly. It is quite possible, also, that the pigs will use this design rather better than the Danish layout as the slightly wider but shorter passage presents a particularly accessible unit.

FLOOR FEEDING WITH OVER-PEN CATWALK

This design has lent itself to further economies in building costs by the elimination of the side feeding passages. This is a very attractive proposition as it reduces the width of the building by some seven feet. Feeding can then be carried out by installing catwalks over the pen at the pen-division height of 3' 6", using a timber-constructed walk (see page 95). A design of this type can be fitted into buildings of many shapes and sizes; for example, a single-sided unit could fit into a building as narrow as 12', with 3' 6" dunging passage and 8' × 9' pens, taking 15 pigs to bacon weight. At the other extreme one could have pens 15' × 6' taking up to 20 pigs, in a building of approximately 34' span. Units of this type are clearly suited in particular to adaptation in odd-shaped buildings, and have especially useful application to those farmers using automated feeding systems of any type.

TOTALLY SLATTED FLOORS

Pioneered in Scandinavia, the fattening piggery with totally slatted floors continues to have a following. Though relatively expensive it is simple to manage and eliminates one of the major problems in slatted dung-passage buildings of persuading the pigs to dung in the right place. An excellent Scottish slatted piggery is shown on page 162. This not only incorporates four sludge channels and pens of modest size (400 pigs in 20 pens), but also has a pressurised ventilation system that moves the air from an overhead duct then over the pigs to under the slats and thence out. This must surely be the best arrangement for such a design. The cost in 1969 was £15.50 per pig.

Totally-enclosed Buildings Using Straw Build-up

So far all the buildings described use little if any straw and, of course, in the slatted-floor type of house no bedding is used at all or the drainage system will not function. There seems no doubt that certain 'behavioural' problems occur in litter-less houses. The incidence of cannibalism, fighting, and tail-biting is abnormally high and causes grave concern to the pig farmer. In most respects this may be said to be a management problem rather than a housing one. Nevertheless, on many farms where the trouble did not occur, and where the pigs had adequate bedding, it has developed since the change-over was made to strawless buildings. The habit seems to develop basically due to discomfort in the pigs, whether it is indigestion or scour, bad ventilation or chilling, overheating or the wrong light intensity. It is therefore more than ever important in a strawless or slatted-floor piggery that the environment and construction of the building be impeccable if good results are to be obtained.

Partly because of this problem, some farmers feel that the use of straw bedding is an important consideration. Also, in many cases there is a need to have straw converted into manure. It is difficult to have this arrangement in a totally-enclosed piggery and generally it is an undesirable practice as the unpleasant and unhealthy evaporations of moisture and ammonia-like gases from the accumulated dung, make the conditions bad for the pigs, the pigman and the building. One design that has had some success, however, is the Hallet type which is basically of Danish layout. By using a wider dung passage of 6' 6", the straw and dung can be cleaned out using a tractor and scraper blade. To give the necessary clearance the dunging area is raised to a minimum eaves height of 7' and is uninsulated.

With this type of house, where the dung is allowed to accumulate over several weeks, and there is a step down between pen and passage of at least 6" to 1', it is desirable to freely ventilate this area; further, if the dung passage roof is raised high enough to give adequate clearance for the tractor with the extra width of the building the whole roof becomes high and the cubic air-space excessive. Thus the dunging passage roof is made as a separate 'lean to' and the space between the eaves of the main piggery roof and the top of the dung passage roof is left open. The ends of the dung passage are also left open so the amount of air circulation in the piggery is considerable.

In order to keep the piggery itself warm and comfortable, a push-in or 'pressurised' ventilation system from the centre of the building is required and also good baffle curtains between the pens and the dunging passages. On exposed sites this could be considered an undesirable design as the air can still overcome these safety measures and cause cold floor draughts. To give more measure of control to the environment in this form of piggery, hinged baffle boards can be placed in the space between the eaves of the piggery roof and the top of the dunging passage roof. It is a disadvantage having to clean out by pushing through from end to end of the unit when the build-up may vary between pens; there is thus much to be said for the individual yard arrangement with gates to each pen and cleaning from the front. The reader is also referred to pages 155 and 158 where the 'Suffolk' and 'Zig-Zag' piggeries are described.

Totally-enclosed Housing for Porkers

The production of quality pork pigs demands an equally high-quality piggery. This need has seen the production of a number of well-insulated units on the prefabricated housing market.

The features of a pork house are largely the same as those for bacon production but with even more emphasis on environmental control and hygiene. It is wise to run the house in 'cell' form either emptying the house, or separate units within a house completely from time to time, as disease problems may easily get on top of the grower. Under these circumstances the house may sometimes be filled during the coldest winter months and the shock of transferring the pigs either from their weaning quarters on the same farm or from another farm altogether, will be very great. It is good practice to minimise this by using whole-house heating for a period with either gas or oil space heaters or an electric fan heater, installing artificial heat at the rate of 250 British Thermal Units per weaner (that is 1 kilowatt per 12 weaners). Economical fan or gas heaters can be purchased for as little as £24 giving 4½ kilowatts which could be sufficient for over 50 pigs. Economy is also achieved by ensuring a high standard of insulation, and a 4″ thickness of mineral wool or the equivalent in other materials completely encircling the building adds remarkably little to the cost yet reduces the heat loss to a minimum.

SLIDING PEN-FRONT

The layout of a porker house will not be fundamentally different from a bacon unit but designs that have achieved considerable popularity are based on the Danish layout, or the inverted Danish with central dunging passage, and have 'sliding' pen-fronts (see page 96). The 'sliding' pen-front ensures that the pigs can be compressed so that they 'fill' the lying area and in this way clean habits are more or less assured. The difference between the area required by the weaner at, say, 40 lb and the porker between 120 and 140 lb is very great: for example, I have seen a pen of weaners lying contentedly in an area of 1·75 square feet each, whereas a porker of 140 lb requires about 4 to 4½ square feet. The sliding front should therefore be capable of reducing the pen size to about one-third of its largest size.

If *ad lib* feeding is practised the fronts may be formed of the metal hoppers themselves; but if floor feeding is practised they may be of metal or wood framing with two skins of ply or asbestos. In recent years there has been a marked tendency towards liquid feeding, and especially with automated systems it is inconvenient to have a movable front. However, the pen can still be kept clean and used economically by placing more pigs in at weaning, thinning out as often as necessary—possibly splitting into two pens later.

In this type of building it is often advantageous to exclude windows altogether and install bulbs of low intensity (15 watt) over the dung passages which remain on all the time, and brighter bulbs (75 watt) for feeding and general attendance over the feeding passage. The lights should be at 10′ intervals in both cases. With porkers it is wise to practise modesty in numbers per pen and a maximum of 20 appears to be quite enough. Such a pen could be 8′ × 10′. With slatted-floor dung passages the pen-fronts should also incorporate gates for weighing the pigs, and entry and exit; there will be no need to have gates along the dung passage. It is very likely from experimental evidence that in practice the *ad lib* feeder is best up to 120 lb and restricted feeding, either on the floor or in troughs, works best thereafter. It is inadvisable to floor-feed earlier than 80 lb, and generally 100–120 lb is better.

The Jordan System

A method of housing pigs that is economical and cuts across many of our preconceived ideas of what constitutes a good environment for pigs has been evolved by Mr J. Jordan of Northern Ireland. As mentioned in the section on environment, the design was originally produced primarily as an aid to relieving respiratory distress in pigs and the secondary result has been that pigs have been found to thrive generally under such conditions.

In Mr Jordan's most popular and copied layout the design consists of a row of buildings, like a row of loose-boxes, closed all round except for a half-heck doorway in front and also a window area alongside the door and under the eaves of approximately 8 square feet. This is fitted with sliding glass panes. No other ventilation is provided. The construction is entirely un-insulated, the roof being of 4″ thick reinforced concrete and the walls of 4½″ uninsulated construction. Under such conditions the condensation is high but this constitutes an essential part of the system. The ideal floor area for this building is 8′ frontage and 16′ depth, making a total floor area of 128 square feet and being sufficient for 20 to 25 pigs. The floor is given a considerable fall from front to back of 6″—and in the front of the pen, running right across the front, is a 2′ wide slatted-floor area with 2′ 6″ wide manure channel underneath (see page 97).

NO FLOOR INSULATION

A further radical departure from normal practice is that there is no floor insulation. The floor indeed serves as a lying/feeding area for meal and, to some extent, a dunging area. A man must go into such a pen daily and clean up as best he can with a squeegee before the meal is fed. Water, whey or skim milk is fed in an automatic bowl or trough, but with the latest system Mr Jordan is feeding on a push-button system to automatic nose-plate troughs, the meal having been agitated with the milk in mixing tanks. It appears that the mortality under this system is in the order of 7 per cent, which is somewhat higher than one might expect with a conventional system even allowing for the fact that all pigs are bought in.

The cost of housing pigs on this system is around a quarter of the cost of, for example, a good Danish house. No system which is capable of growing 23,000 pigs a year, as Mr Jordan's does, can be lightly dismissed. It must be emphasised that a system like this requires the greatest possible skill in management. Many of those who have tried it have failed; reports exist of pigs being killed by heatstroke, by poor control of ventilation; in the Jordan establishment up to 21 air changes per hour are a common air-change rate in winter, which corresponds to a very favourable rate of 12 cfm per pig. It is thus not an ill-ventilated house, but careful control of the window and top-door openings must be made to ensure these conditions are maintained. Those who have copied the system but have failed have, in fact, neglected important points like this and those who are considering this arrangement should study the Welfare Codes given in Appendix 6 which deprecate any condition which could cause distress—as indeed will any pig farmer. The problem with the Jordan system is that the margin between comfort and discomfort is a narrow one.

PEN SIZE

A fundamental question in the design of fattening quarters is how many pigs should be penned together and how we should grade the pens to make maximum use of the area. The general concensus of opinion is that groups of fatteners are best in lots of not more than 15–20, with 10 perhaps as the ideal. It must, nevertheless, be stressed that some farmers can rear up to 40 together with apparent success; such concerns are those with a high and unique standard of stock management.

It is essential that when the pigs are lying down in the pen they more or less cover the floor—otherwise dirty habits will develop and muck will be deposited in the pen. The problem arises as to how one can ensure this when a weaner will occupy only about 2 square feet of floor space when recumbent, whereas a baconer occupies some 5 square feet and a heavy pig $5\frac{1}{2}$ to 6 square feet. Several solutions can be offered. One is to design a pen with a sliding front so that the area can be enlarged as the pigs grow. This system means that when the pigs go in the pen as weaners they can carry right through to finishing in the same pen and in the same group, which many will consider an advantage, minimising the check that occurs when pigs are moved and reducing the possibilities of cannibalism and fighting. It is particularly suited to the short growth period of the porker. An interesting alternative approach is to have a totally-slatted floor (page 99), suited to porkers or fatteners of any age and referred to on page 138.

Another solution for baconers and 'heavies' is to have pens of two sizes, one for the growing stage from weaning to, say, 16 weeks and the finishing pens from 16 weeks (100 lb) to finishing. If it is desired to have 15 pigs to a pen, the area of the grower pen would be 8' × 6' (48 sq ft) and the finishing pens could be 12' × 6'. This arrangement envisages *ad lib* or floor feeding in the grower stage and floor feeding in the finishing stage. If troughs were inserted, 12 to 13 pigs only could be penned under this arrangement. For bacon production two finishing pens would be needed for every grower pen.

Yet another arrangement is to have a weaner pool at eight weeks in which young pigs are placed in fairly large pens, 20 to 30 to a unit. They can be allowed 2 to 3 square feet of lying area and kept there until 100 to 120 lb. At this stage, the best ten are taken off to the finishing pens, leaving the remaining number for a few more days when they can be divided off into well-balanced groups in the finishing pens. One weaner pool pen may therefore serve three finishing pens. It is likely that the mixing of several litters at weaning creates a 'stress' from which the pigs may take some time to recover under intensive conditions and it is for this reason that the deeply-bedded yard with warm kennel lying area is more popular, allowing up to 8 sq ft area per pig. The same system can be used if multiple suckling is practised, but without the severe weaning stress of several changes at once.

I have already referred to accommodation for early weaned pigs on pages 129 and 130, and it will be appreciated that the various forms of housing referred to may be used for a sufficient

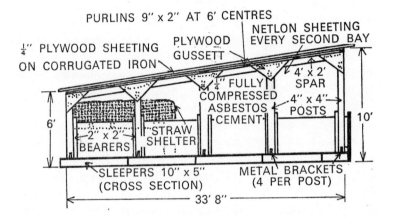

PURLINS 9" x 2" AT 6' CENTRES

NETLON SHEETING
EVERY SECOND BAY

PLYWOOD
GUSSET

¼" PLYWOOD SHEETING
ON CORRUGATED IRON

¼" FULLY
COMPRESSED
ASBESTOS
CEMENT

4' x 2'
SPAR

4" x 4"
POSTS

10'

6'

2" x 2"
BEARERS

STRAW
SHELTER

SLEEPERS 10" x 5"
(CROSS SECTION)

METAL BRACKETS
(4 PER POST)

33' 8"

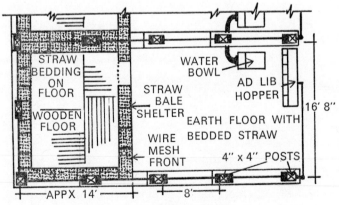

STRAW
BEDDING
ON
FLOOR

WOODEN
FLOOR

STRAW
BALE
SHELTER

WIRE
MESH
FRONT

WATER
BOWL

AD LIB
HOPPER

EARTH FLOOR WITH
BEDDED STRAW

4" x 4" POSTS

16' 8"

APPX 14'

8'

Fig. 25. *Weaner pool with straw-bale kennel and covered run.*

145

length of time to allow the pigs to go from their pens, cages or verandah units direct to the finishing house. There is certainly no absolute rule in this respect but it should certainly be a main endeavour to reduce the number of changes in the housing to a minimum.

A simpler arrangement is to have one size of pen only, stock this as full as possible at weaning and then, as the pigs grow, 'thin out' some as porkers, letting the best only carry on to bacon or heavy weight. This is a rather 'laissez-faire' arrangement and not particularly commendable.

Whenever pigs do muck in the pens it is always a help to clean and disinfect the floor and then place a barrier across part of the pen so that when they are laying down they really do fill the available space. The great limitation on the shape of a pen is that with pigs being trough-fed, 9" to 15" of linear space has to be allowed to each pig, so that pens are usually long and narrow and also the building itself. With floor feeding such limitations are removed and much more economical buildings and conversions can be made, but as mentioned previously floor feeding is often unsuccessful up to about 100 lb liveweight.

DUNG DISPOSAL: ENCLOSED PIGGERY

In the totally-enclosed piggery there are various ways of dung disposal. With solid-floored passages, the erection is cheap and the dung may be disposed of in solid or semi-solid form. Bedding may be used or not, but frequent cleaning is necessary as it cannot be considered satisfactory to allow a build-up of dung inside a totally-enclosed piggery as the evaporation of moisture and ammonia will make the internal atmosphere very unpleasant and the effect on the building structure undesirable.

The traditional method for a Danish-type piggery with solid-floored, side dunging passages is to clean out with shovel and barrow and provide trapped drains to take off excess liquid. With such a design the dung is in solid form and bedding is optional. This method of cleaning can be mechanised by using a small mechanical horticultural cultivator with blade attachment and by pushing through from end to end. Whilst this can be used with side dunging passages, it is easier with the arrangement of a central dunging passage and pens on either side so that there is one movement only of the machine through the piggery.

An alternative is to have a solid-floor dunging passage with a step down from pen to passage of 3″ to 4″, omit the drains, but have a virtually flat floor so that the passage can be cleaned out with either a mechanical scraper or with a squeegee that is made to exactly fit the width of the passage. This is used in piggeries without bedding or with very little, and produces an end-product in slurry form that can be dealt with as a liquid. An alternative disposal system with a solid-floor dunging passage is to have a slit along the outside of the dunging passage which falls into a drainage channel or sludge pit under the passage. It is not entirely self-cleaning even though a fall of some 2″ will be provided in the direction of the 3″ slit.

The simplest arrangement for dung disposal is the slatted floor with automatic drain or sludge pit underneath. Apart from its fully automatic nature, it is also probably the most hygienic as the dung and urine pass to the channel below and have no inter-pen contact. Clean pens, however, must be maintained and research has shown that pens approximately twice as deep as they are wide favourably influence this necessity. In addition there should be a slatted area at least 3 ft or more in depth and approximately 6 sq ft of uninterrupted floor area for each bacon pig. Finally, the house temperature should not exceed 70°F if possible.* Very considerable experience and development on slats for pigs has been gained and published by Soutar, and the recommendations below are largely his (see also diagrams on page 105).

The slats themselves may be constructed of several materials; concrete, metal or welded wiremesh. Reinforced concrete is probably the most popular and wiremesh the cheapest. Suitable

*Partially Slatted Floors and Floor Feeding in Pig Fattening Houses. Livingston, H. R. and Robertson, A. M. Ex. Farm Build. Rep. No. 8, N.I.A.E., Silsoe, Beds.

dimensions for all types of pigs are a slat width of 2″ to 3″ at the top tapering to 1½″ to 2″ at the base, a depth of 2½″ to 3″ and a gap of ⅞″ between the slats. Metal T-bar slats available in proprietary form are made 1″ × 1″ × ³⁄₁₆″ and spaced alternately ¾″ and ⅜″ apart, and supported by 2″ × ⅛″ flat. Welded mesh is best using 3″ × ½″ mesh at 10 gauge or 3″ × ⅝″ mesh at 5 gauge. The last method though the cheapest is not always found to be satisfactory. For pigs three months and over 6″ × ¾″ mesh, 5 gauge, is satisfactory. Costs vary considerably, ranging from 50p a square foot for T-bars, to 30p to 35p for concrete and 20p to 25p for weldmesh, including supports.

The sludge may be allowed to accumulate under the slats or slit, and be pumped out periodically through drainpipes fitted at intervals and opening to the outside. The pipe must be fitted with a cap to prevent draughts blowing into the pen.

In calculating the volume required for sludge, one should allow for approximately 3 cu ft of sludge per pig per week with whey feeding (which produces the maximum), down to 1½ to 2 cu ft per pig per week with meal feeding. If lengthy storage facilities are needed, the cost of the sludge tank will be considerable and one of the main disadvantages of this system.

Alternatively, a drain may be placed under the slats to take the slurry to a tank at the end, or ends, of the building. This channel is usually 2′ deep, 3′ wide or less at the top, sloping to 1′ across at the base, which may, if desired, have a half-round glazed pipe (see page 105). A fall between 1:120 and 1:80 is required and should not go outside this range.

At the end of the channel where it enters the tank, a sluice-gate of metal in wood runners is provided so that the sludge may be periodically run off into the tank. Some piggeries, however, do not provide a sluice-gate and it seems to work just as effectively with sludge trickling slowly into the pit. If the sludge at any time fails to flow correctly a 40-gallon drum of water can be emptied at the top end to push through the muck.

The sludge in the tank can be dealt with in a number of ways. It may be pumped out from the collecting tank and distributed through an irrigation system, or it may be pumped into a vacuum tank and spread over the fields from this. For these two latter processes quite costly equipment is needed and also the cost of the tank is considerable. Access to the fields must be maintained for much of the year. An alternative way of dealing with the effluent is to run it on to straw or deep-litter from poultry broiler or laying houses and then spread it in solid form; this commends itself to many farmers who are unable to deal with the sludge in liquid form yet desire the benefit of the slatted-floor system in piggery management.

FLOOR FALLS
The fall on a solid-floored dunging passage to the drain is normally 1 in 20 to 1 in 30.

Fattening Piggeries With Yards
Whilst the totally-enclosed fattening house represents the most popular trend at present and provides designs of general application throughout the country, under some circumstances piggeries with outside or covered yards may be advocated. The advantages of such designs are several. Generally a 'yarded' piggery is cheaper; the section of the building which needs to be well insulated is of more modest proportions and so the cost of materials and erection can be reduced. The yard itself can be left uncovered, which is a particularly cheap form, or can be covered with a single skin of asbestos sheeting or other protective sheeting material. Also, some will consider the environment obtained by such designs basically healthier, for the dung and urine are outside the warm part of the building. Ventilation measures can therefore be on simpler lines. Generally speaking uncovered yards are most suited to the West and South-west of this country where the climate is milder and frosts infrequent or less severe.

TIGHTLY-PACKED PIGS
The pigs are packed tightly in the insulated area, the atmosphere is free of ammonia and other fumes evaporating from the dung, and warmth is easily retained so that the high-tempera-

149

ture conditions required are readily available. The lesson from Sweden (referred to in chapter 2) certainly indicates that this may be one way of dealing with pneumonic diseases and alleviating symptoms for the 'cold' piggeries at Sala, Sweden, are essentially no more than covered yards without even the luxury of a warm kennel and yet here pigs are healthier than in some indoor accommodation, and utilise their food reasonably well. The explanation may be that pigs in small groups in a healthy environment may have such a 'positive' bill of health that they can still outdo in food efficiency their unhealthy brethren kept in warmer accommodation.

An important feature of this type of housing, and indeed all forms of housing incorporating outside yards, is that if any dung build-up is intended it must be *outside* the insulated and warm part of the house. Notable designs were conceived and produced between 1935 and 1945 in which the dung build-up was permitted within the area warmed by the pigs and these proved virtually unworkable, as with even normal ventilation rates the evaporation from the dung and urine caused an unpleasant environment and severe condensation and deterioration to the building. If ventilation was increased to prevent these undesirable effects, the piggery would be too cold for much of the year.

The greatest problem that has always had to be faced by the farmer with outside yards is how to keep them warm and free of draughts in the cold weather. For this reason the size of the individual unit within the building must be kept small and each unit must be completely separate from the other units to prevent through-draught. Also, the siting and aspect of the building are critical factors. The yards should be of southerly aspect and face only in this direction; a double-sided open-yarded piggery is undesirable because not only may the wind blow through from one side to another, but the yard facing east or north will always be relatively cold. Good baffle arrangements between the yard and the pen are necessary, and a reverse-acting ventilation system which blows the air into the piggery and out through the pop-holes is desirable, to prevent the danger of floor draughts.

A Single-Sided Yarded Piggery

We may take as our pattern of the yarded piggery a good design of a single-sided deep-straw piggery with covered yard and individual cleaning. It is based in broad outline on the best features of McGuckian design from Northern Ireland and W. T. Price's Harper Adams Pig Parlour. The design, shown on page 103, consists of a service passage running along one side of the unit, the wall containing double-glazed dead-lights. Off from this passageway would be catwalks at right angles serving pens on either side, each one being 16' in length and 6' deep assuming a trough is used.

There must be doors along the service passage between each pair of pens to prevent draughts blowing in the pop-hole at one end and out at the other. This part of the building is well insulated and should be ventilated by installing the push-in ventilation system either by a duct running in the angle between the roof and the north wall with the fans at each end, or the duct may run at a central point in the pen immediately below the roof or above a false ceiling where needed. An alternative is to place fans in the north wall and blow air into the pens, with a baffle placed in front to protect the pigs from direct draught. Both the baffles and the ducts must be double-lined and well insulated with vapour-sealed material. The walls between each pair of pens are taken to the roof—in fact, they may be load-bearing and support it.

The yard is separated from the pen by a cavity wall. A pop-hole not more than 2' 6" wide and 3' 6" high allows access for the pigs. Above the pop-hole may be a half-heck door to make for easier access between the pen and the yard, facilitating such operations as the weighing of the pigs.

BAFFLING THE DOORWAY

The extent of the baffling of the doorway is going to depend to some extent on the siting and locality of the piggery. In reasonably sheltered situations, with full southerly aspect and the 'push-in' ventilation system referred to, it should only be necessary to have a good 'curtain' at the pop-hole (see chapter 13).

In more exposed situations, however, a full baffle-system is needed with small porch outside to produce a form of air-lock between pen and yard. The depth and extent of the yard will depend on the amount of straw it is required to build up. The length of the yard must be limited to keep the covered area to economical proportions and the building as compact as possible, whilst the width will only be some 7' with feeding passages arranged in this way. A 2' drop to the yard, with ramp for access by the pigs, will give a build-up of 3–4 months, depending on the amount of straw used, whilst a 1' drop will last about 6–8 weeks. In the latter case no ramp need be used.

The gate on the front must be strong and is best of galvanised metal which at the base should be treated with a good bitumastic preparation from time to time, whilst inside this, to the height of the dung, should be muck-retaining boards to keep extreme pressure off the gate. These are simply planks fitting into slots at the side, immediately behind the gate. The gates swing outwards and the yard is cleaned with tractor and foreloader. This makes it essential that there is a fair clearance between ground level and the eaves. A satisfactory distance is in the order of 9'. If a lean-to type roof is used, the building will become rather high at the north side. It will therefore be better to have a pitched roof, in practice, and install a flat false ceiling over the living area to conserve heat.

This form of building may be said to represent the perfect arrangement, for this type of piggery. The pigs are penned in small numbers, in separate units with easy stock management and inspection and good environmental control.

The original Harper Adams design was of basically similar conception but the troughs ran alongside the service passage, the yard was uncovered and there was no step down into the yard. Consequently the building tended to be long and narrow and the straw usage was heavy; in any event an uncovered yard may become very wet in the winter and the muck is trodden back into the pens so it is difficult to keep the pigs clean. Also, the feeding passage—without doors between the pens—leads to draughts at floor level and of an uncontrollable nature. It is

Modern feeding piggery with yards; doors between pens isolate groups, cut down draughts and aid disease control.

Face-to-face range of fattening pens on the late Mr. Stephen Horvat's farm at Worlingworth, Suffolk.

Group of fatteners in a Trobridge fattening house, where they are floor fed and have no straw.

The pigs in the Trobridge fattening house are fed their pellets through the roof flaps at the rear of the building.

Lidded sleeping pens in a Solari-type fattening house.

Left: Dunging yard in the Solari house, showing steps to allow for dung building up. *Below:* Outside view of a Solari house in Suffolk.

Sow house and yard, with boar pen and yard at end; individual sow-feeding stalls in front.

Individual feeders for sows in covered yard.

therefore essential to have doorways down the passage and also to have a particularly efficient baffle porch between the yards and the pen in this design. If a 'push-in' ventilation system is not used, there should at least be a controllable chimney outlet to each pen to provide individual ventilation.

THE GARCEAU 'SQUATTIE'

A still simpler 'yarded' piggery—in fact, probably the simplest of all—is the Garceau 'Squattie' which is used for young pigs *ad lib* fed. It employs very low-roofed and uninsulated kennels only three concrete blocks high. These kennels are 7' × 5', giving a total area of 35 square feet to take about twelve weaners. The kennel is situated at the back of a yard 14' deep and 8' wide, and along the back as well as the kennel is the self-feeder. The wall of the yard is only two blocks high and is surmounted by an electric fence—otherwise the pigs would easily jump over a wall 1' 8" high! (see page 101.)

The low kennel and walls are a deliberate and successful attempt to ensure a draught-free kennel, as the wind will blow right over the kennel and yard and the low walls generally ensure no eddies and back-draught into the kennel.

The design was originated in Cornwall and is undoubtedly most suitable for the milder southern-most parts of the country. In the colder regions the system can hardly be advocated as the snow, the frost and the wind would make the control of the environment for the pigs and management well-nigh impossible at some times of the year. In the southern part of the country it may be considered a reasonably satisfactory method of having a cheap form of intermediate housing between weaning and the final controlled-feeding stage of fattening, though it does have the disadvantage that with this arrangement the pigs are lacking a controlled environment when they most need it in the fattening stages.

The Garceau house represents a modern adaptation of the cottager's pig sty. There are, however, other advocates of the cottager's sty. For example, a simple range of pens leading to uncovered yards is often used for whey-fed pigs in southern areas. Whilst the control of the conditions is not complete, with cheap and highly nutritious feeding pigs will thrive. There is no doubt that highly profitable enterprises exist, feeding pigs in large numbers in this way. The whey and meal feeding are both in the yard outside, the muck may be squeegeed and/or hosed into a drain to run to a sludge tank or the outside yard may be slatted (see page 163) and the pigs lie in warm insulated kennels behind the yard. Keeping pigs in relatively small self-contained units always tends to healthier stock so that one sees the same advantages taking place here as in the Swedish 'cold' piggeries and healthy pigs, well managed and very cheaply housed are not easily pushed into second place when it comes to profitability! Further great merit can be attributed to the cleanliness in these units with the yards frequently cleaned and washed down.

PUSH-THROUGH CLEANING

It may be desired with solid floored yards that mucking out be carried out more easily and frequently by running a tractor straight through the yards with scraper-blade attachment. This is effected by providing gates between the yards that swing back, enclosing the pigs in the pens and giving a clear run for the tractor. Such buildings were popular a few year ago when all totally-enclosed buildings had a bad reputation following the failure of the uninsulated Danish-type buildings.

DEEP-DUNG DESIGN

An alternative approach to this problem is to have deep straw, allowing some build-up of muck, with kennels at the back for the pigs to lie in. In this design the troughs are at the front of the yard and feeding is from a passageway in front with overhang in the roof to protect the pigman and troughs. The cleaning-out is carried out by swinging the gates back to the kennels and pushing through; it is clearly most satisfactory in such piggeries to clean out fairly frequently to avoid too great a depth of yard and make overall construction difficult.

An original design of this form was produced by the late Mr

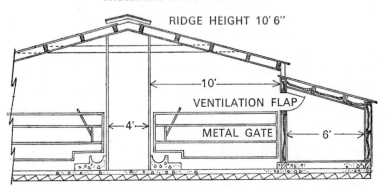

Fig. 26. Suffolk-type fattening house, with insulated kennels, covered yard and push-through mucking out.

Stephen Horvat of Suffolk (see page 98). A refinement which is now quite popular and is referred to as the 'Suffolk' house (fig. 24) encloses the centre area completely but this is uninsulated and naturally-ventilated copiously. The kennels at the back are very well insulated and have flaps to control the air supply. Plenty of straw must be used in the yard and cleaning out should be fairly frequent to ensure a healthy atmosphere. It should be accepted that environmental control will often be less than perfect. In addition great care is necessary in controlling the ventilation flaps to prevent chilling the pigs and consequent coughing and pneumonia. The design has become popular with those farmers who have found no favour with more intensive, and especially slatted-floor and litterless, units. Another type is shown on page 161 (Fig. 29).

The merits of such a system as this are that it is cheap and also keeps the pigs in small groups quite separate and in warm cosy kennels. Against this, any fine environmental control is lacking and the kennels may well become too hot and stuffy at some times of the year. Under these circumstances the pigs may lie in the yard much of the time and avoid the kennels, which is neither good for management nor health.

DOUBLE-SIDED UNIT

I mentioned earlier in this chapter that with yarded piggeries the aim should be to have single-sided units with yards facing only in a southerly direction. The only circumstances when it is permissible to have a double-sided unit is if the site is a particularly well-sheltered one. Also, the two sides must be completely separated to make sure there is no possibility of wind blowing through the unit from side to side. Where there is a central feeding passage it is best if it is divided by a longitudinal wall to ensure that no through-draughts result.

Yarded Piggeries of Adaptable Form

There are a number of designs of fattening piggeries with yards that aim to make the building adaptable for other stock as well as pigs. Whether this is a wise choice is open to question as a piggery is a building that requires considerable specialisation if it is to work efficiently—and the more adaptable it is for other stock, the less suitable it usually is for pigs.

Adaptability in pig housing probably represents an outmoded idea for a modern highly-specialised industry. One example of the adaptable building is seen in the Eastern Counties in which the pigs are kept in a large covered yard. The layout, as shown in the diagram on page 98, has a side service passage with troughs off at right angles, similar to our first design (page 103). Insulation of the sleeping area is provided by storage of the straw above the sleeping area. The sleeping area is raised well above the dunging area which allows a large build-up of dung. In some cases the drop is as much as four feet between the pen and the dunging area and this allows the build-up to be for as much as a year. Gates between the pens in the yarded area are lifted out and the tractor with fore-loader can clean through quite easily.

A suitable arrangement of this type is a house 37' wide. With side service passage of 3' and pens of 16' and yards of similar size, the building is at least very adaptable and houses the pigs cheaply. It also stores straw and provides for only a periodical clean-out. The commonest design has troughs along the service passage as well as the catwalk to provide for the maximum number of pigs per pen. This would, in fact, make the number up to 30 pigs per pen, which is rather too large.

USES A LOT OF STRAW

A design of this type uses a great deal of straw and is most suitable for large arable farms; it has not generally been found too satisfactory or an ideal way of fattening pigs but it obviously may fit into a particular system of farming and provides a building which goes some way to being convertible.

SOLARI FATTENING HOUSE

One of the undesirable features of a building of the type described above is that there are really too many pigs per pen. The Solari design overcomes this criticism by placing under its open-yarded roof a raised feeding passage with 'kennels' on both sides, each taking ten pigs. The kennels have 3' 6" high brick walls and hinged 1" thick boarded tops which may be hinged or slide open. It is by these that the ventilation is controlled. The yards are 18" below the pens and gates at the side open out for cleaning by tractors and muck-loader. With a 3' centre passage, and 7' wide pens on each side and 6' wide dung yard, the total span of the building is 30' (see page 100).

In the right hands the Solari fattening piggery has given good results, but the keeping of pigs in kennels rarely seems to produce the very best results, and individual attention required by the pens is considerable.

Several adaptations of this design have been produced as the broad arrangement of the Solari house has had a large following. The cost is certainly rather more modest than a traditional totally-enclosed piggery with troughs, probably varying from three-quarters to two-thirds of the cost. The possibility of adaptability also commends itself, though the cost to convert it would be considerable and wasteful with the dismantling of the fittings, gates and so on. Many consider the environment is healthy for pigs, but apart from the isolation of the pigs into small groups, there is nothing particularly healthy about this housing system and the conditions in the kennels are a comparatively unknown quantity.

Frequently nowadays the 'Suffolk' type of house referred to on p. 155 is built in the form of kennels against the side of simple yarded structures similar to the Solari pen. This is illustrated on p. 163. It is perhaps preferable to have the feeding in the dunging and exercising area, in which case all forms of feeding are easily dealt with from the central feeding passage.

Another very popular but rather similar alternative is the 'Zig-Zag' house with kennels in the centre of the general-purpose building giving access alternatively to feeding and dunging areas twice the width of each kennel. Both these designs use straw, and also store it above the kennels, and clearing is daily or two or three times a week by tractor and scraper.

There are quite a number of alternative arrangements even for this selection and probably mistakenly some farmers have used these designs in much wider buildings, using four or more rows of kennels. One of the most modern kennel arrangements is the 'kennelled Ulster" developed by Mr Malcolm Priest. In this, seen on page 160, the troughs, for automatic liquid feeding, are in the kennelled area and the dunging area is tractor cleared and on the outside of them. With 17 pigs per pen it costs somewhat less than the 'Suffolk' or the 'Zig-Zag.'

SLATTED FLOOR IN PLACE OF YARD

Judging by the complaints of respiratory and other diseases in kennel accommodation, stress in the widest sense of the word can be an important factor. One adaptation is to replace the yard with a slatted floor; this is, of course, particularly suited to farms where little straw is available. The dung may accumulate in a pit under the slats or be taken to a tank at the side or at the end. Alternatively, and even more simply, the dung yard may be replaced by an ordinary solid-floored dunging passage of Danish conception; the dunging slit (see section on slats) is sometimes used here and is a popular form in Ireland.

Floor feeding can be practised in these kennels as readily as in other piggeries. Weighing will usually be done in the centre passage, where there is no special weighing room, and so gates must be provided in the front of the pen to allow movement of the pigs. The pigman has reasonable protection for his work but certainly not of the same standing as in a totally-enclosed house.

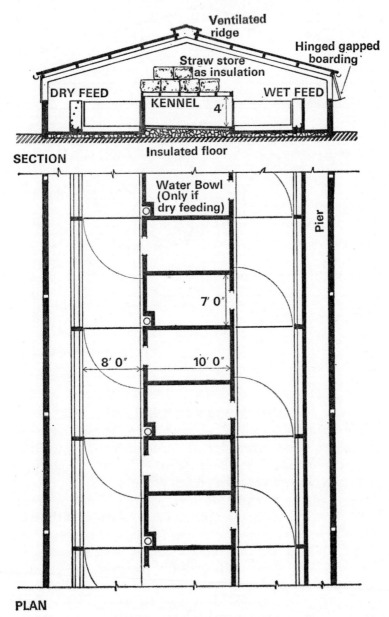

Fig. 27. Zig-Zag fattening house.

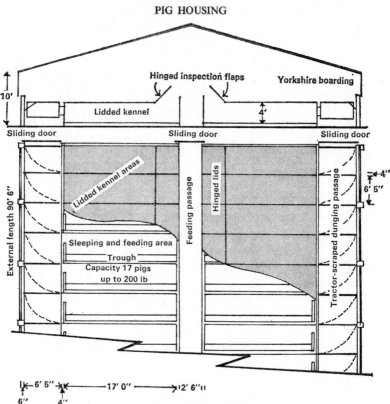

Fig. 28. *Kennelled Ulster fattening house.*

Open-fronted Fattening House

The merits of several of the more 'open' housing forms are exemplified in the Trobridge open-fronted houses (page 102). The pens are 15' × 8', should face south and take 20 to 22 porkers or 14 to 15 baconers. Each pen is completely isolated from its fellows on each side, has a sharply sloping insulated roof and overhang and flap in front, and ventilator at the rear. The pigs lie in an area demarcated for them at the back by a board, which is normally bedded and the front dunging area 4'–5' wide is solid, or may be slatted if preferred. Feeding may be in *ad lib* hoppers at the front, or on the floor from the back, or liquid feed into troughs along the sides. The system is economical in cost and generally healthy, but management, and especially of the ventilation, must be diligent and knowledgeable if results are to be good.

160

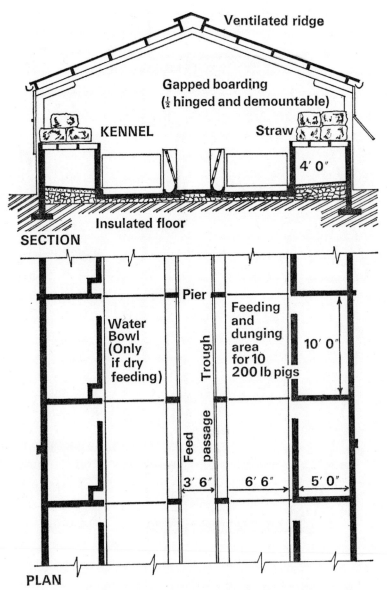

Fig. 29. *Suffolk-type piggery in a covered yard.*

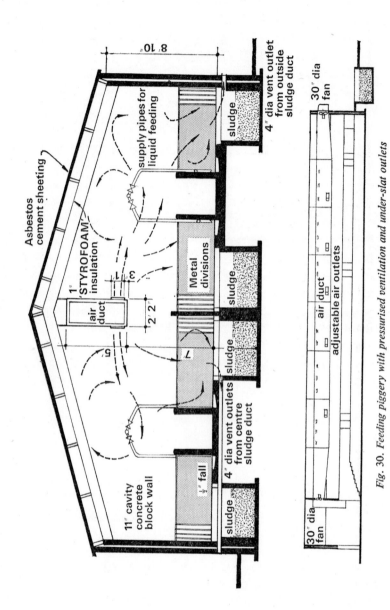

Fig. 30. Feeding piggery with pressurised ventilation and under-slat outlets

162

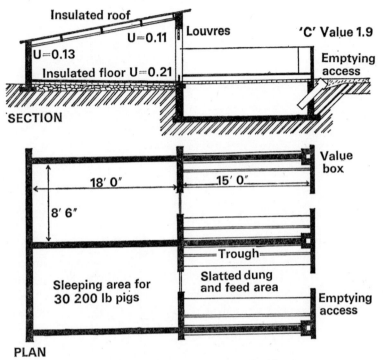

Fig. 31. *Cottage-type piggery with open feeding/dunging area.*

163

Chapter 10

ACCOMMODATION FOR SOWS AND BOARS

IN many respects a disproportionate amount of attention is given to the housing of fatteners in comparison with breeding stock. Though it is certainly true that under some circumstances there may be no real harm in breeding stock roughing it out-of-doors, and in fact they may appear to flourish under such conditions, the inevitable tendency and need for more intensive methods of keeping pigs means that correct housing becomes more vital. It is often alarming to see the scanty attention that breeding pigs receive and the filthy conditions under which they are kept. Their accommodation need not be expensive to be good, but it does require more forethought than is usual.

The Fully-covered Yard

An excellent way—indeed perhaps the best way from the pig's point of view—of housing dry and in-pig sows is to keep them in a completely-covered yard. Such a yard can provide complete protection from the weather and under these circumstances the amount of straw used need not be prohibitive. And whilst the system enables the sows to be kept under the healthiest and most invigorating type of conditions, it keeps them well protected from bad weather, allows plenty of exercise and makes the provision of good stockmanship easy. Individual feeders can be provided and a suggested layout is shown on page 106. The system is based on a line of feeders down one side of the yard which are raised above the general level of the yard itself—the further the raised area the greater the build-up. A drop of 2' 6" will allow a build-up of approximately 3–4 months.

A suitable yard on these lines can be provided by having a span of 30' and dividing it into bays of 15' along its length. This gives a total area of 450 square feet per bay, which is suitable for eight sows. This is an ideally small number to keep together as the likelihood of fighting and bullying rises as the numbers kept together increase. Along one side the individual feeders have a length of 7' including the trough, so that the actual lying and exercising area is just over 40 square feet per sow, which can be considered a satisfactory and generous one.

EASILY REMOVABLE GATES

Gates and fences between the sections of the yard should be easily removable to allow for easy cleaning. The construction of the yard will be on simple, uninsulated lines but extra comfort can be provided by storing much of the straw on a platform above the strawed area, thus also making much good use of the yard. Alternatively, the straw can be stored to one side of the yard under a lean-to extension. In both cases the straw can be thrown down onto the floor without difficulty.

The ventilation of the yard is an important aspect that requires some careful attention. A yard of this type can be closed in on three sides, that is, the back and the two ends; the ridge must be left ventilated, either by using a capped open ridge, or by installing chimney-type ventilating trunks. The front of the yard should face south and it is best to have some protection here from the elements; the best way of doing this is to have an overhang of 5' on the roof; this will give the pigman protection in feeding and also protect the yard from undue entry of snow or rain. If there is no overhang, then the area over the feeders should have some form of top-hinged shutters that can be brought down in bad weather. A minimum height to eaves of

10' can be considered essential to allow clearance for the muck loader.

The base of the yard should be of concrete as not only will this assist in its cleaning, but it will also prevent the dangers of disease build-up which can always be a serious problem where ground is used continuously to house livestock. The base of the walls of the yard can be of solid brick or concrete construction and the remainder of the wall above this to the eaves of the yard can be in corrugated asbestos or iron, or can be very well constructed in slatted boarding. This consists of 4" boards with $\frac{1}{2}$" gaps and provides a good measure of ventilation without draught and entry of rain and snow.

Partly-covered Yard

Good as the covered yard is, it does represent a relatively expensive way of housing breeders; a cheaper way of dealing with this matter, and probably little inferior in practice, is to have only part of the yard strawed and covered and the remainder composed of a concreted area partly for exercising and partly for the feeding, containing the usual individual feeders. The simplest layout would be similar to that in the totally-covered yard, with the lying area at the back in the form of small kennels, allowing 10 square feet of lying area per sow, and a reasonably generous concreted area in front of 30 square feet per sow. At the far side of the unit will be the individual feeders served by a concrete apron.

With this design the sleeping den can be very modestly constructed with an overall height of 5' to 6'. Whilst it is clearly advantageous to have full cavity wall and floor construction (as in a totally-enclosed piggery) it is not essential, but a damp-proof construction is, and whether bricks or blocks are used the exterior should be rendered or treated to keep it dry. The roof will need some insulation to prevent condensation and undue temperature fluctuations; a simple lining of fibreboard is just sufficient, though when one considers the small cost in inserting a vapour-sealed mat of mineral wool behind it, it is clearly a worthwhile proposition, or even simpler is the use of polyurethane sheet. To keep the sows dry and warm there should be a step-up into the pen of a few inches and a small sill; the entrance need be only 2' 6" wide and half-heck doors can be used with the top kept closed in the cold weather. There is usually no need for further ventilation except for a sliding shutter at the back of the pen, measuring 2' 6" × 1' for a unit of eight sows.

The diagram shows a modest-sized unit suitable for ten sows (see page 106). Units of this type can be expanded to be as long as required, with gates between the ten-sow unit so that a tractor can clean through quite easily. An alternative and useful layout is to range two units side by side and have a feeding passage between serving sow feeders on either side. Which one is selected depends on the site and the eventual expansion desired.

Some breeders are concerned with the number of leg and foot troubles that occur when sows are kept on built-up litter and they may actually prefer this latter design as the exercising on hard concrete may produce a healthier reaction on legs and feet. Our own experience with keeping sows on built-up litter has not, in fact, been unsatisfactory and the feet have kept in good trim. Troubles are perhaps more likely to occur if insufficient straw is used and there is occurrence of "foot-rot"-like organisms which build up in the soil and the dung. The trouble may therefore be more connected with management than housing.

Sow Stalls

When sow stalls were introduced a few years ago they represented a revolutionary concept of sow housing. Stalls are similar in size to a farrowing crate and allow the sow only to stand up and lie down; she can neither turn nor exercise. The idea originated in Scandinavia and in these countries the practice has been to tether the sow in a stall, like a cow-stall, with collar and chain round the neck, a trough in front and with timber or metal partitioning. The stalls are 2' 6" wide, 3' high and 6' 7" long, with two retaining chains behind.

Sows can be kept in this accommodation from the time they are weaned from their piglets to the time they return to the farrowing pen, the only time they leave being for service. Alternatively, they may go into this accommodation after being safely in-pig, the previous "dry" weeks being spent in a yard or paddock adjoining the boar accommodation.

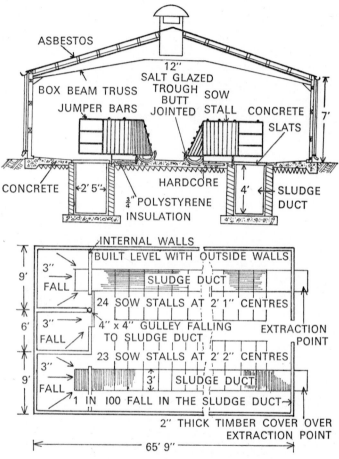

Fig. 32. Sow stall house, with boar pens on left.

At first sight such a system might appear to be at variance with many of those concepts of good sow management we have accepted. Experience shows, however, that most traditions in pig management are to be questioned and with nutrition suitably adjusted the sow-stall system can prove satisfactory. It can

General view of Harper Adams "Natural"
farrowing huts, designed by W. T. Price.

Farrowing and rearing quarters on a Scottish
pig farm

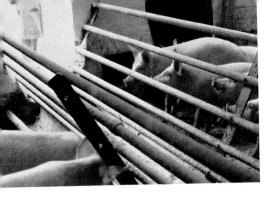

McGuckian-type troughs and pen-fronts with feed catwalk between the two lines.

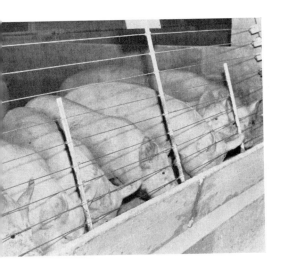

Trough-fronts with steel wire.

Swinging pen-front keeps fatteners from the trough until their feed is ready.

Tubular individual sow stalls on a Scottish farm.

Voluntary stalls for pregnant sows.

The Harper Adams "Crark"—an outdoor hut designed by John Luscombe.

The "Pigloo"—a proprietary form of sow shelter.

Improvised sow shelter—three bends of corrugated-iron sheeting.

be an economical form of housing as the stalls may be placed in simple narrow buildings some 14' wide, though it is more usual to have them in two or more rows. The system also has the advantage that it ensures each sow has a fair share of food, freedom from fighting and bullying throughout pregnancy, and uniformly equable conditions.

The Scandinavian system of tying the sows by the neck has not found as much favour here as in other countries, but nevertheless some large units have been erected and have met with success. A suitable design is shown on page 107. The short, tubular stall divisions are cheap and the open area at the back allows the sows considerable movement and makes cleaning out as easy as possible with a solid floor. Bedding may be used if desired. There is also a system of girth-tethering which is preferred by some since it allows a greater degree of movement.

The more usual arrangements are for untethered sows and there are several basically different designs. Originally solid-sided stalls were most popular with the sides reaching the floor and being 28"–30" wide, but if a 6" space is left betweeen the side and the floor the stall width may be reduced by 4". Tubular-steel divisions are much more popular and are certainly advised and may consist of horizontal rails only or a combination of horizontal and vertical rails (see page 89). Stall fronts are also normally railed and occasionally fitted with gates to allow easier movement of the sow. Rear gates may be solid, which is best when a partially-slatted floor is used to retain any surplus dung, or when bedding is used with a solid floor. The doors may slide vertically or be hinged. With solid-floor stalls vertically-sliding tubular gates or chains are used, the former being designed to clear the floor so cleaning out is facilitated.

The length of stalls if a raised trough is used is 6' 6" from the rear gate to the front edge of the trough, the trough itself forming an extra 1' 3", but if a sunken floor trough is used, a total length of 6' 9" is sufficient.

Raised troughs, required for wet feeding, should be of half-round glazed fireclay not less than 12" wide, but floor troughs, 12" wide and 2" deep are otherwise preferable and can be lined with tiles. The sow lies with her head over them. *Ad lib* drinkers are usually installed, with one drinker being shared by two sows. It is best to have the water level controlled by a tank and float-valve at the end of the row to minimise the working parts. Nipple drinkers are also frequently used.

Floors may be solid or the rear part slatted. The solid floor should be well insulated and non-slip, the front half being level, and the rear half sloped well back with a fall between 1″ and 2″. With a slatted rear end the front half falls the same amount to the rear 3′ which is slatted. A good slat is concrete, $2\frac{1}{2}″$–3″ top width and $\frac{7}{8}″$–1″ gap. The slats are best laid at right angles to the divisions. An invaluable detailed study of stalls has been made by Seaton Baxter and is reported in *Stalls for Dry and In-pig Sows*, Scottish Farm Buildings Investigation Unit, No. 3, Jan. 1967.

Sows kept in stalls should be able to see their fellows and if two rows are used they should be housed face-to-face rather than back-to-back as in the latter case the sows strain to see what is going on behind them. The sow-stall house itself must be very well insulated and ventilated, bearing in mind that often the sow will have no bedding, she will be fed frugally under modern techniques and there will be no heat generated from exercising. Extremes of temperature and dampness will there-fore be potentially especially harmful.

At present there is still considerable development taking place in the search for the ideal housing for the sow. Sow stalls have certainly enabled individual attention to be given to the sows, but some people do not agree with the restriction on the movement of the sow and also object to the unclean condition of the sows' hindquarters in litter-less pens.

Two interesting approaches have been evolved which keep sows in small groups. Sow cubicles, which were pioneered by Bill Marshall of BOCM, consist of a group of three or four 7′ long × 2′ wide free-choice cubicles for feeding and lying with a communal dunging area behind them of 6′ × 6′. The unit would normally be placed under a covered yard and with gates between each dunging area could be mechanically cleaned.

Cubicles may also be placed in an outside hut with separate or communal dunging areas; one design has a door that can be opened and closed by the sow. These are all useful approaches with the same ends as all dry-sow husbandry—to give a stress-free environment at minimum cost and with easy management.

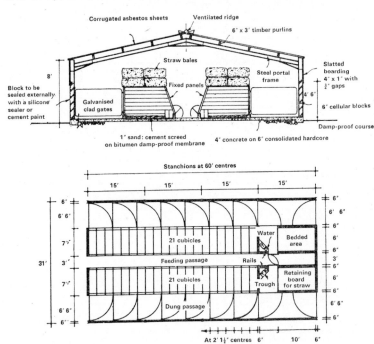

Fig. 33. Sow cubicle system developed by BOCM Ltd. Boars are housed in bedded areas.

Housing the Boar

'The boar is half the herd.' From the genetical standpoint this is true enough and this emphasises his importance. The correct form of housing is also important because it prolongs his existence and use in the herd and aids his fertility. Boars are apt to 'go off their legs' and suffer from a number of mechanical troubles of the legs and limbs which may be less likely with good housing. From the health and vigour point of view probably the best way of keeping the boar is outside in a paddock with the

171

simple protection of a hut and run. Where more confined accommodation is required, a simple system of sty and run is all that is required but the exercising area should be on generous lines. A covered area of around 40 square feet and yard of not less than 80 square feet should be the aim. Part of the yarded area can form the service area or crate which should be designed so both the boars and sows to be served can be easily moved around.

The type of unit shown in the diagram (page 107) can be installed in the middle or the end of a range of sow yards. Another satisfactory and indeed very good way of keeping boars is to run them with a group of sows safely in-pig. No special accommodation is required in such cases—in fact a boar can be run with a group of up to eight sows under any of the other systems shown. Whatever system is used it is generally an advantage to have the boar within sight, sound and smell of the weaned sows awaiting service as there is evidence that this can have a beneficial effect in stimulating the signs of heat. Where sows are kept in stalls it is usual to walk the boar behind the sows to aid the detection of service.

Accommodating the Newly-weaned Sow with Boars

Because of the need to get sows served successfully as soon as possible after weaning and since it is most likely to be achieved if they are kept close to the boar—in sight, smell and sound—it is preferable to give special consideration to this period.

Most of the accommodation I have considered in this chapter incorporates boar units close to the sows or gilts, and a design is outlined on page 106 that will help to make the process of detecting heat and serving the females, an easier and less laborious job.

Farmers often prefer to place sows in this type of accommodation for about 4 weeks after weaning before they enter cubicles stalls or other system for the last 3 months of pregnancy.

Chapter 11

OUTDOOR HOUSING

APART from farrowing huts which we have already dealt with, outdoor accommodation for other pigs is also required including the dry and in-pig sow and the young gilt or boar. For fatteners outside accommodation is virtually never justified due to the large amount of energy that would be wasted in their excessive movement—probably as much as 30 per cent—and also the unavoidable difficulties with labour and management.

Outdoor accommodation may always be simple and one of the most satisfactory ways of building warm and comfortable outdoor huts is by using a relatively temporary erection of straw bales (see page 108). A hut two bales high is usually sufficient—each bale being 3' 6" × 1' 6" × 1' 10". If the bales are three tiers at the front and two at the back a good fall is provided on the corrugated-iron roof. With internal dimensions of 10' × 7', there is room for eight adult pigs quite comfortably. The roof should be packed underneath with straw held in place by wire mesh to prevent condensation and excessive temperature fluctuations. To preserve the hut the lower part of the walls may be protected with corrugated iron or strong wire mesh. A usual practice is to destroy the hut after a few years when it may become parasite-ridden.

A design has also been produced of straw bales erected on and around a 'Dexion' iron framework that can be moved across the ground drawn by a tractor, thus overcoming the only disadvantage of the straw-baled hut that it is immovable so that there is a temptation to over-use and contaminate the pasture ground associated with it. 'Permanent Pastures Perpetuate Parasites'—and so also would straw-huts that were used indefinitely!

CURVED CORRUGATED SHELTERS

An even more simple form of hut is one made of curved corrugated-iron sheets. A few of these sheets erected on dry ground and closed at one end only form a cheap and simple hut. Each sow should be allowed approximately 8 square feet of lying space.

In the more luxury class several shapes of huts can be used. Simple curved roof accommodation can be made in corrugated-iron or exterior-grade plywood. A hut of length 8', width 6' and height 4' can be easily moved on skids by tractor or even by hand, being comparatively light. One end is completely closed, whilst the other has a small doorway. Apex huts, similar to the farrowing houses already described, are also used but without the refinements and apparatus needed for farrowing. Single-skin construction is always sufficient for adults.

LEAN-TO SHEDS

Lean-to sheds are still commonly used and a popular size for these is 7' × 5', 4' high in front and 3' behind. At a floor area of 35 square feet four sows could be accommodated in comfort. Very popular second-hand forms of accommodation are 'retired' insulated railway trucks which are cheap and extremely warm and dry. They are, however, heavy to move and are probably most satisfactory used as accommodation pulled up on hard standings. This is, of course, a common way to use all forms of hut accommodation in winter when the management of pigs kept outdoors may be difficult or the land may be quite unsuitable.

The floors of the straw or corrugated-iron huts can normally be left as bare earth provided that they are on a dry, well-drained area. If not, timber floors are strongly advised. A liberal covering of straw should always be provided. Ventilation in accommodation such as these huts does not seem to require any specific alteration provided that there is a good opening at one end. This may be partly covered during the winter with a sacking curtain or top-hung door, and left fully open during the summer. Huts should always be faced away from the prevailing winds.

Chapter 12

ADAPTATIONS AND IMPROVEMENTS

A LTHOUGH in converting an existing building into a piggery much may depend finally on the individual circumstances of the case, a number of general principles can be stated that can usefully be applied to conversions in general. The same can be said of those piggeries that require improvement; experience shows that difficulties usually lie in certain well-defined directions. In this chapter I have therefore listed the main points to look for and the possibilities presented in the conversion of buildings of the most commonly-used dimensions.

Insulation. Usually the roof is high and uninsulated; in such cases a false ceiling at working height—say, 6′ 6″—is the best compromise, reducing both the cubic air space to be warmed by the pigs, and the heat loss. A good ceiling is made by an oil-tempered hardboard inner lining with a 2″ thickness of polythene-sealed mineral wool on top. A rather cheaper alternative is to use an appropriate polyurethane or polystyrene board (see page 52). But if these are considered too costly, 2″ of straw may be suspended by wire mesh, inserting a vapour-seal of sisalkraft or polythene above the wire and below the straw.

The walls may also need attention including extra insulation and damp-sealing. Existing single-thickness brick or block walls, or those of uninsulated concrete blocks, may be insulated by placing an additional thickness of lightweight concrete blocks on the inside and rendering at least up to a height of 4'. A cavity should be maintained between the existing and the new wall. A further method of dealing with this problem is to fix battens to the inside of the wall and fix on these 1" or 2" thick wood-wool slabs (which are then rendered with a cement and sand screed) or $\frac{3}{8}$" or $\frac{1}{2}$" asbestos sheets with vapour-sealed insulation behind. To prevent moisture coming through the wall from the outside, the best treatment is to have a waterproof rendering, but as this is expensive it is acceptable, but less permanent, to treat the outside with two coats of cement paint or with a silicone water-repellant preparation.

DEALING WITH WINDOWS

Windows are often a source of heat loss and condensation—particularly when a large area is involved as in many old buildings. It is usually quite satisfactory to place an insulation layer on the inside of the glass, such as polystyrene, and then cover it with an asbestos or oil-tempered hardboard cladding.

Floors. Existing floors are generally of dense concrete and are unsuitable for pigs unless it is intended to use copious amounts of straw bedding. To correct this usual fault an insulated layer is placed on top of the existing floor after a damp-proof course has been inserted underneath the new insulation. The damp-proof course and the insulating layer can consist of the usual materials, but a good choice for the latter is a 1" thick wood-wool slab with $1\frac{1}{2}$" screed over, all to be set to the new falls that will probably be required. The extra covering will also have the effect of raising the floor above the dunging area, so giving the desirable 'step-down' into the dunging passage that is normally required.

Ventilation. The usual difficulties arise from uncontrollable open ridges, as in cowsheds, or a lack of any stale air outlet at all. The procedure for extraction of stale air is to close off any uncontrollable ventilation and insert a chimney-type extractor from the false ceiling through to the ridge. The use of mechanical ventilation can very often be strongly recommended as it means less interference with the basic structure of the building. It may

M

be possible to make one small opening, assisted by an efficient fan, serve as an extractor from the piggery, whereas with only natural ventilation several openings would have to be made. Inlets of the usual type, mentioned in the section on ventilation, must also be provided, adequate in size and distribution if draughts and condensation are to be avoided.

DIVISION OF BUILDING

The arrangement of pens within the building will now principally depend on the width. A few examples will make the possibilities apparent.

Width of Building	Farrowing and Rearing	Fattening
12′ approx.	Single row of pens. 10′ × 8 ′ and 4′ service passage	Single row of pens. 3′ dunging passage. 3′ service passage. Pens 10′ × 6′ with or without troughs
20′ approx.	Double row of pens. 10′ × 8′ and 4′ central passage	Double row of pens. 4′ central dunging passage. Floor feeding by catwalk over pens or trolley over dunging passage. Pen size 8′ × 8′
32′ approx.	Three rows of pens with two parallel service passages. Pens 10′ × 8′. Passages 4′	With trough feeding: pens 6′ × 10′, troughs running between pens. Dunging passages on outside 3′ 6″. Service passage 3′ 6″ central
45′ approx.	Divide building into two longitudinally. Have two separate double rows of pens as the 20′ span house	With trough feeding: have central service passage 5′ wide and side dunging passages 5′ wide. Pens on either side to measure 6′ × 15′, with troughs and catwalks between pens, to take 15–16 pigs. With floor feeding: pens of similar size will take nearly 20 pigs

Chapter 13

FIXED EQUIPMENT

The Catwalk Arrangement

WHERE the feeding passage is at right angles to the main service passage no trolley has to be taken along it and the width can be much reduced. The McGuckian-type trough arrangement under these circumstances is ideal and is also one of the cheapest to install. It is shown in fig. 1, page 109. It consists of a 1′ wide passageway set at a level with the top of the troughs on either side. The front consists of 2″ × 2″ × ¼″ angle-iron with rods, threaded through, of ¾″. The second rod from the bottom is carried on a piece of angle-iron which is set back into the pen to avoid food falling on to this rod.

This type of catwalk will be set above the service passage, one step up, if the floor of the pens is on a level with the main service passage floor. If, however, the service and feeding passages are on the same level feeding is rather easier. The pens will then be some 6″ below the passage levels. With automatic wet feeding which is now so popular the arrangement of two troughs backing on to each other between pens can be simplified as no catwalk is needed, the tap being controlled at one end outside the pen. It is, however, extremely important to prevent fouling and wastage and the diagram (page 186) shows the design evolved by David Taylor. The troughs are raised and have a rail above to prevent the pigs lying in them.

179

Gates

Some of the noticeably weak features in piggeries are the gates—and particularly to the pens, dung passages and yards. The rooting and destructive capabilities of the pig must never be underestimated. Satisfactory gates can be made of a number of materials. Metal is undoubtedly the best material but it should be well galvanised or the deterioration will be very marked. Timber gates are also satisfactory but should be protected with sheet metal at any point at which the pig can make contact. Recently gates of $\frac{1}{2}''$ thick asbestos have been marketed which promise a corrosion-free material. The gate must also be well secured on its hanging side; it gives the strongest fitting if it hangs on a steel post or channel independent of the dividing wall. Likewise, the gate must be closed with a latch that is secure, pig-proof and still easy to operate, preferably with one hand.

There are a number of good designs on the market of pro-prietary type—suitable and well-tried forms are shown in the diagrams (figs. 2 and 3, page 109). The greatest strength is needed in yard gates which have a build-up of dung on one side —not only this but they need careful protection against the eroding effects of the urine. Timber gates are rather heavy for this purpose and the ideal is metal, well galvanised, and pro-tected periodically with black bitumastic paint to prevent decay.

In many cases it is common practice to see gates placed along the line of the trough. This wastes valuable trough space and is unnecessary. We have overcome this difficulty most satis-factorily by making a sliding gate over the trough of small-diameter pipes which telescopes inside the larger tubes forming the trough front. To get the pigs out of the pen, a ramp of inverted-V shape with as gradual a slope as possible, and slats to give the pigs a foothold, is placed over the trough. Pigs rapidly learn to use it; only the first time is much persuasion needed!

Baffled Doorways

One of the problems repeatedly emphasised is that of draughts blowing from open or covered yards into the pens themselves. In order to minimise this, several devices have been used of baffle curtains or self-closing doors (fig. 6, page 111). An ordinary sack, soaked in sheep dip first of all to discourage the pigs biting it, is quite successful but lasts only a limited time; an alternative is a rubber flap. Apart from manufactured products, discarded rubber conveyor belting is a cheap answer. It can be obtained from the National Coal Board at nominal cost.

So far as self-closing doors are concerned, a top-pivoted doorway, of light wood construction protected with metal, which swings in either direction is reasonably successful but may still allow a considerable amount of air to come in around its necessarily loosely-fitting sides. It is also easily prevented from closing by an accumulation of dung or straw. An alternative is a side-hung door which is set at an angle so that it is also self-closing. This opens outwards only. The pigs will push this type of door from the inside easily, or when moving into the pen from the outside they will quite readily 'edge it' outwards with their snouts.

One or two excellent self-closing double doorways of proprietary manufacture are made in metal with adjustments allowable for various sizes of pigs. Whatever system is used it is essential to maintain it regularly and keep it in good working order. Further assistance is usefully obtained in preventing floor draughts by arranging a step at the doorway—this may consist of a central 12″ high step with 6″ high step-ups on each side so that pigs may easily walk up and over it at all ages.

The Weighing Room

An essential to the study of progress in fatteners is constant weighing, at weekly or fortnightly intervals. As methods of feeding and dung cleaning become more efficient, weighing becomes one of the most time-consuming chores in the fattening house and every effort has to be made to speed up and make this chore more efficient.

In most piggeries the arrangement is to take a portable weighing machine round the piggery pen by pen. But this has several unsatisfactory features; the machine is neither highly accurate nor well damped and the recorder cannot take the figures under the easiest conditions. A more satisfactory arrangement is to

have a fixed weighing machine in a separate place with arrangements to ease the task of the recorder and the stockman, and
also to allow quick circulation of pigs.

In most piggeries the arrangement is to take round the piggery pen by pen a portable weighing-machine but this has
several unsatisfactory features; the machine is neither highly
accurate nor well damped and the recorder cannot take the
figures under the easiest conditions. A more satisfactory
arrangement is to have a fixed weighing machine in a separate
place with arrangements to ease the task of the recorder and
the stockman, and also to allow quick circulation of pigs.

A good design is a semi-circular arrangement with swinging
division (as below) A plan of the operation would be to
bring the pigs in from door 'A' with swing division back in
position 'X'. As the pigs pass through the weighing-machine
the swing division is pushed round until at position 'Z' the last
pig goes through and the pen of weighed pigs extends again to
door 'A' for their return to the pens.

If pigs are weighed in the pen or passage attached to the pen,
careful thought must always be given to the direction in which

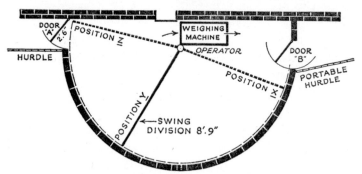

Fig. 34. *Semi-circular arrangement with swinging division for weighing pigs.*

the doors open or close to arrange correct holding and circulation of the pigs. Wherever possible it is better to weigh in a
feeding passage or the pen rather than in the dunging passage
which makes conditions most difficult. When fixed machines
are used a pit can be arranged so that the platform is on a level
with the floor; such an arrangement makes the pig far more

willing to walk on to it than if he has to walk up a ramp, even if slight, which is not rigidly stable.

Loading Pens

It is a very risky procedure to allow the lorries collecting the pigs for market to come into close proximity with the rest of the pig unit, due to the great dangers of bringing in disease. To prevent this, a passageway should be provided from the piggery to a loading bay holding the maximum number of pigs that are likely to be sent off at one time. A further help is to provide a raised loading bay to allow the pigs to walk on level with the lorry. If this whole set-up is near the highway but at least 100 feet from the fattening house, the danger of disease introduction is virtually eliminated and the job becomes an easy one.

Self-feeders

With self-feeders it is wise to allow three pigs to a foot of feeder space. Self-feeders of wood or metal are satisfactory and may be used as part of the pen divisions with economy. A useful procedure is to incorporate the self-feeders in the pen-fronts; these can slide forward or backward to give the variable size to a pen which is a great help in ensuring cleanliness in the stock.

A simple type of self-feeder, built into a pen, is shown in fig. 8, page 111. Most self-feeders can be filled automatically by chain-and-flight or overhead conveyors.

Individual Feeders

It is strongly recommended to provide individual feeders in sow yards to ensure fair shares and prevent bullying. The usual type of feeder is of metal construction—measuring 1' 10" across and being 6' 6" long. The gate at the back for the sow should be remotely controlled from the front which will also contain a built-in trough (see page 106).

Water-bowls

A necessary permanent fitting in the piggery is the automatic water-bowl. As self-filling bowls are generally used and there is always some spillage, it is most satisfactory to place them in the dunging area; where this is impossible they should at least be situated at the lowest point in the pen adjoining the dunging area (see fig. 9, page 112).

The bowl is best placed well within the passage so the pig has its whole body in the passage when drinking. It is also important that the bowl does not project into the passage when it has a solid floor to interfere with cleaning-out operations. To satisfy this requirement the bowl may be either placed on the dung-passage door, connected by flexible piping, or recessed into the dividing wall between pen and passageway. The bowl lip should be 6″ above floor level, but where young pigs before weaning are using it, it is good practice to place a step up to it. This keeps it cleaner and less likely to be fouled.

An important consideration is that the bowl should not be placed in an area where frost may affect it. Care must be taken when it is in an outside yard, therefore, that the bowl is as near the pen as possible and pipes are kept warm either by running within the pen or by efficient lagging. It is usual to have a storage tank to serve the bowls rather than direct from the mains; if this tank is placed within the piggery it should be covered and lagged to keep the water inside clean and dust-free, and also to prevent condensation which will drip on the floor and wet it.

The nozzle drinker has in recent years achieved a large measure of popularity, the water flowing when the pigs depress a valve on the end of a brass nozzle projecting from the wall, gate or pen division. The system is cheap, hygienic and should give little mechanical trouble, but it is wise to select a good one rather than a cheap one as many have a short life and soon 'drip' or run excessively. Pigs are also rather inclined to play with them and waste water.

There is some evidence that if the water is warmed in winter that pigs will benefit from it; a tank within a piggery will ensure the water has the chill off it. Alternatively, bowls can be purchased which incorporate small heating units.

Where restricted water is given rather than *ad lib* to save physical handling of the water, the bottom rail over the trough is, in effect, a water pipe. This pipe is individually controlled by a valve to each pen and is punctured on the base by a series of

Sows in these stalls, fitted in a converted Dutch barn, are held by a 2-in tether round the girth.

Farrowing crates in a walled-off part of the Dutch barn unit, shown above, in which the sows are also girth-tethered.

Central dunging passage of cubicle system for pregnant sows, developed by BOCM Ltd.

Below: Sows, grouped in threes, lie in individual stalls but can stretch their legs in the communal dunging passage.

Piglets on this trials farm enter these cages after weaning at 7-10 days. They are dry fed and drink from an automatic waterer.

Double radiators in front of air inlets in this cage-rearing set-up raise air temperature to 72°F. Electric heaters (left) boost temperature to 80°F in the first stage of rearing.

These flat-deck cages with wire floors can be used both for 3-week weaning and as a second stage for battery-reared weaners.

Below: Interior of circular piggery, showing pen arrangement and feeding boom.

small apertures ⅛″ diameter at 9″ centres.

Troughs

When fixed troughs are used in the piggery great care should be given to the small details in their design and construction, particularly bearing in mind the importance of eliminating food wastage at all costs—food representing 80 per cent of the cost of rearing pigs. The general requirements are that they hold the necessary food and allow the pigs ready access at all ages within the group, yet provide as little wastage as possible. A trough must also be hygienic, easily cleaned and able to withstand acidic elements. Their arrangement is to enable the pigman to put the food in easily without wastage from his side, and also without undue interference from the pigs. Trough space allowances should be:

    6″ at weaning,
    9″ for small porkers,
    10″ for heavy porkers,
    12″ for baconers,
    14″ for heavy hogs.

Undoubtedly the best trough is made of half-round salt-

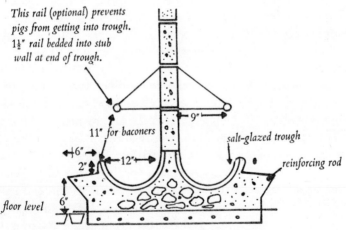

Fig. 35. Taymix design non-fouling trough.

glazed pipe. Though this is expensive it satisfies all the requirements for the trough itself that have been specified above and
there seems to be no substitute that is as satisfactory. Ten-inch
or 12-inch diameter pipes are used, the latter being preferred.
Fig. 7, page 111, shows the detail of this with the tubular-rail
trough-front. The trough is tilted towards the pig and should
be 5″ above the floor on the pen side while the front should
extend 1′ 3″ above the feeding passage. The lip at the top of
the trough nearest to the pigs should be a maximum of 2″ wide.
The rails over the trough-front must be set accurately in the
way shown to ensure, firstly, that pigs cannot escape, and
secondly, that food is easily put into the pen. A recess under
the trough-front, as shown in the diagram, should be 1½″ to
2″ to help the pigs to stand right up to the front.

ADVANTAGES OF RAISED TROUGHS

Certain evidence has been given in America and Sweden that
extra ham development may result from raising the trough about
18″ above the floor and providing a step up for the front feet.
A device such as this both for troughs and water bowls will
certainly lead to cleaner management and a minimal chance of
pollution of either of these fittings.

The tubular-rail front is the most popular and cheapest and
being fixed rigidly does not 'wear' like so many other things in
the piggery. There are, however, advantages in the 'swinging'
trough front, the chief being that the food may be put in the
trough without interference from the stock and whenever it is
convenient prior to the actual feeding-time. This can represent
a great boon to the pigman at week-ends when all that need be
done is opening of the trough, if necessary, by unskilled labour.

The swinging-front arrangement is shown in fig. 10, page 112.
Whilst there are several good proprietary forms made in metal,
several points need emphasis whether proprietary or home-made
types are being used. The design of the trough is different as it
must be set square in the concrete with its lips level and with the
2″ steel-tube suspending rod for the front being set over the
centre. A swinging front should not extend more than 10′ in

length to ensure rigidity and it is best made of light metal—solid sheeting for preference, tubular or welded mesh as an alternative. Timber fronts are used but are heavy and rather difficult to operate. The gate should be fixed by a locking bar which is centrally operated and fits rigidly at both ends—both being critical points if the fronts are to be easily operated and the pigs are not going to damage them by their rooting and shaking.

A cheap alternative to the tubular rail front is the use of high tensile wire strands as developed by Mr M. J. Parker (see page 104). Use of this method for a number of years has shown its durability and success.

Dividing Walls

Walls between pens for fatteners need not be more than 3′ 6″ high; for farrowing pens 4′ is advised. Walls between the dunging passage and the pen are normally the same, or in the case of slatted-floor arrangements they are either omitted or a dwarf wall only is used. I am inclined to think, from the problems that occur in persuading pigs to use a dunging passage, that it is preferable to have a clearly demarcated area by using the full-size pen wall. It is helpful with slatted-floor arrangements where there are no gates along the dunging passage to have two openings into it at each corner of the pen. This will be a further stimulant to the pigs to use the dunging passage as it will eliminate the 'dead' corner of the pen which they are most likely to foul.

Wherever possible, divisions between pens and between pens and dunging passage are taken to the ceiling. Ceiling-high pen divisions taken at regular intervals across a large piggery, together with doorways between the individual compartments of the piggery so formed, do have the most favourable effect on health and on the environment.

Service Room

The pigman must not be neglected. Not only does he need a place to keep protective clothing and boots, but he needs a place where he can take and collate the many records required in an efficient unit. This involves no more than a small, well-

187

insulated room, ideally attached to the farrowing house, with lighting and heating points.

Single Stalls for Fattening

A problem that has always existed has been the difficulty of finishing the slower growing pig after the main group has gone. To leave them in a pen is a thoroughly wasteful procedure, and to mix them with other pigs is likely to be most harmful. Perhaps the best answer is to have a few pens with individual stalls, which are best on slats, where these 'odds and ends' can be finished. Units are now commercially available and measure 5′ long and 1′ 3″ wide.

Automatic Feeding Systems

Considerable attention has been paid to automated feeding systems and the pig farmer now has the choice of several to fit in with most systems of husbandry. These have been mentioned where appropriate in the text but it may be helpful to summarise the possibilities here. Troughs may be filled mechanically, by travelling hoppers, with meal or cubes, or dispensers above the troughs or *ad lib* feeders can be filled with chain-and-flight conveyors, or augers. The same principles can be used to fill dispensers suspended above the floor for automatic floor feeding. The dispensers themselves after filling may be released by hand or automatically, whichever is preferred.

Automatic liquid feeding into troughs is probably now more popular than dry-feed arrangements for fatteners and there are several manufacturers supplying equipment of varying complexity. All systems have a main mixing tank where the constituents can be automatically delivered and then mixed with water or milk by-products. From here the liquid mixture is pumped round to the houses or pens and the mixture may be released into troughs by individual control of taps at each point, or more sophisticated electronic arrangements have a central control board whereby the quantities to each trough can be accurately regulated from a central place. Other arrangements to give accurate delivery to the pens depend on valves controlling the period of delivery.

Automated systems require good management to ensure uniform and hygienic feed supply and provision must be made for the whole system to be cleared through when necessary. Clearly a disadvantage is that it is normally only possible to dispense one ration to the group of pigs but this seems more serious in theory than in practice. Undoubtedly automation in feeding will be essential in the future.

Chapter 14

FUTURE TRENDS IN PIG HOUSING

IN the preceding chapters in this book I have looked at the present-day position in pig housing and tried to give an up-to-date resumé of current practice and present developments. To conclude it may be profitable to take a look into the future and see what developments may lie ahead in this field. The next few years certainly promise to be an exciting period and an era when we may anticipate more efficient designs at a lower cost per pig than at present. Building costs *per se* are almost certain to rise but it is because they have been relatively cheap in the past, even with outmoded traditional methods of building, that we have tended to be so extravagant with the floor space and the cubic areas of our piggeries. We have also, until very recently, devoted too little thought to the labour involved in operating our piggeries. A great volume of pioneering work is proceeding at the moment and the benefits from this will surely be felt within the next year or two.

DISEASE, HUSBANDRY AND HOUSING

It is inevitable that a further growth of large intensively-managed units will take place, where the operation of looking after the stock can be attended to most economically and the buildings and services are so centralised that costs are kept to a minimum. Both in this country and elsewhere attempts have been made to organise co-operatives, whereby the pigs are bred on a number of specialised farms and then moved to one or two central fattening units for finishing. This type of enterprise requires a particularly high degree of skill to organise without running into serious disease problems. Traffic in pigs between sites is to be avoided as far as possible and a more satisfactory solution would be to build up a number of completely self-contained units where all the operations are carried out on one site, and the overall size of the units is not excessive. New genetic material must, of course, be introduced, but with proper quarantine or by using artificial insemination the risks are greatly reduced. For example, a unit of 70 breeding sows with followers, producing some 1,200 fatteners per annum, would make a satisfactory enterprise for one man and an assistant. This would require the following accommodation, numbers being approximate and depending on the intended output. The pattern presented here would serve as a guide to the establishment of any self-contained unit of this size, with the necessary multiplication as the size was increased. I would suggest that a maximum for one site should be about 280 sows.

Breeding

10 farrowing pens split into two units

20 rearing pens, also split into two separate units

or, alternatively, up to 30 combined farrowing and rearing pens in three sections. The divisions are essential to enable this disease-vulnerable section to be well rested and disinfected between batches.

Yard accommodation to be provided for 40 sows in units of not more than 10, but preferably as few as 6. There would be no excessive expense in providing 7 units of 6 in totally or partially-covered yards. Where possible, the sows would run out on pasture. Sow stalls or cubicles would be a suitable way of housing the dry stock.

Four boar pens would also be needed.

Fattening

Fattening accommodation for approximately 400 would be

required if baconers are to be produced. Any of the appropriate modern designs of piggery could be selected hereafter, dealing with such questions as type of pig required, the acreage available for dung disposal, the method to be employed for the removal of sludge and dung, and whether food is to be in solid or liquid form.

Houses for an enterprise such as this can be grouped most economically and movement minimised in the following way. The fattening unit represents the end of the 'conveyor line' so this should be nearest to the road to enable easy loading from a ramp, which should be placed at a distance from the house so that the truck does not come into the confines of the farm. With a central service road for the pig unit running at right angles to the main road, the two fattening units can lie on either side nearest the main road. Moving back along the road, next would come the rearing units, then the farrowing quarters and finally the sow yards and boar pens. This arrangement would keep the traffic of pigs to a minimum and also keep the breeding stock as far inside the farm as possible. Another advantage would be that the sows could be easily moved on to the fields, and the muck from the yards is also at the right end of the unit for disposal.

SITING THE FOOD STORE

Another consideration is the food store; on farms where milling and mixing is carried out in a central store, a small holding place only is needed at the centre of the pig enterprise, a building of 200 sq ft being large enough. Where the food is coming in from outside, however, a store adjoining the fattening houses would be best, the food being in a position where the majority is used, and also not being too far distant from the other units.

Other buildings that may be needed include a weighing room, lean-to accommodation for implements, and a Dutch barn for straw. This latter building could be sited at the north end, to shield the lower pig buildings from cold winds.

MINIMAL-DISEASE UNITS

I have described above an example of how an enterprise may be developed to cope with the specialised production of 1,200 fatteners per annum. In general the development of a number of modest sized self-contained units, well scattered, will help in disease control, The temptation towards the very large unit on one site is great with the inevitable pressures from the economist favouring this arrangement, but it should be resisted because of the disease dangers.

The unit for the 'minimal disease' pigs, which are initially free of most or all of the common pig diseases, could be on the lines of my suggestion above. Additional precautions will, however, be necessary. A complete double fencing should surround the unit and no one must enter without changing into a complete set of protective clothing that is kept on the site.

With a fully-intensive unit it is feasible, and probably desirable, to place a cover over the yards to keep out the birds, as well as protecting the totally-enclosed houses from bird entry. Food and litter may bring in disease so the trucks must only back up to the site from an arranged position, the food and fodder being thrown down into the unit itself without entering it. Every precaution must also be taken in the design of the pig-houses to make them vermin-proof.

STANDARDISATION AND PREFABRICATION

Some progress must and surely will be made towards standardisation. Farmers still want many different designs to suit their own ideas, but such individualism makes it extremely difficult to reduce costs. The urgent need is the development of a limited number of standard designs so that the shell of the building at any rate can be prefabricated and can incorporate a very high standard of insulation. Within this framework a certain amount of flexibility is quite possible, but farmers must not be such individualists that everything has to be different or they will hurt themselves and keep the costs unnecessarily high. Prefabricated building manufacturers look for quantity as an essential in reducing their costs. Let us also finally lay the ghost of prefabrication being synonymous with flimsiness. This is a

193

N

hangover from the immediate post-war years when a low standard of prefabricated housing was used, many of which, incidentally, are still standing. Prefabrication merely means doing as much of the construction in the factory as possible, where it can be executed properly by skilled hands. It generally entails a more satisfactory construction than traditional building and the use of specialists in the design and execution.

The use of the highest standards of vapour-sealed insulation, mechanised ventilation and so on are not procedures that can be reliably left to the local builder, and the use of good specialists in prefabrication relieves the farmer of much worry. In the author's own unit at Cambridge University, this method of construction has been used for every type of pig building and has not only produced successful construction, but has reduced the building costs by about one-third. It has, incidentally, given us thermal insulation standards for the walls, roof and floor, nearly three times higher than those associated with traditional construction. Also, all interior surfaces at pig level may be cement-rendered or of equivalent strength and hygiene so that the highest levels of cleanliness may be maintained.

FEEDING

It seems clear that for fatteners above about 100 lb liveweight the trough can be recognised as being unnecessary where dry feeding is practised. This fact has already had its effect on design but it will have a more profound effect as more farmers gain confidence in its use. With floor feeding and particularly the use of an overhead catwalk, considerable economies are possible, and this may have further effects on pig-housing design. With, however, the experimental evidence now available to show advantages in liquid feeding, automatic liquid feeding into troughs is certain to have its attractions on the larger holdings. In most cases on the smaller unit the pig farmer will prefer to be with the pigs when they are fed and deliver the feed himself, as this is the time when he is able to study the behaviour of the pigs and see that they are feeding correctly.

The best procedure at present seems to be to mechanise in every way possible the conveyance of the food to the site, with the actual delivery still in the hands of the stockman. Excellent designs of mechanically-operated feeding trolleys are now available which can deliver predetermined quantities of food into pens in the easiest way, thus easing feed delivery while still retaining the stockman's presence.

DUNG CLEARANCE

The path is now open for eliminating all labour in dung clearing from piggeries. There is no doubt that slatted floors with automatic drains underneath, to take the dung away in slurry form, represent a certain choice for the future where the farmer can make use of this type of manure.

In other cases it can be handled by a dung scraper in solid form. Mucking out is a chore without merit and every piggery should be built so as to make this as easy a task as possible. Cultivators with bulldozer blades or tractors with fore-loaders can be used for dung passages or strawed yards, and the range of equipment within this category is so wide that almost any design can be mechanised in one way or another.

CLIMATE CONTROL

In the pig-house of the future, climate control will be achieved as far as possible by automatic means. Natural ventilation is quite feasible in many designs, particularly when the unit is small, as in the 'individual cell' piggery such as the 'Squattie' or numerous kennel designs. However, because labour is needed for its constant adjustment, natural ventilation may be expected to give way increasingly to fan ventilation controlled semi-automatically with fans and thermostats. Thus, the job of controlling day-to-day fluctuations is made easy and stable temperatures are ensured. Where mechanical systems are employed, the fresh air supply will be taken firstly to the pigs, then over the dunging area and thence to the outside. In this way actual ventilation needs will be minimised and both the pigs and the pigman will work under ideal conditions. Modern systems in the future are unlikely to neglect the great importance of giving the operator completely favourable conditions; the stockman is

N*

the most important inhabitant of a pig-house and he must have the environment and the facilities to do a skilled job.

The piggery's climatic needs have been clearly stated; it is unlikely that any great advances will be discovered in this field at a practical level in which so much work has already been done. Temperatures in the order of 65° to 70°F will be maintained in intensive fattening houses, but greater store will be placed on maintaining constancy and uniformity of temperature and other conditions than on the actual thermometric scale. Ventilation systems are unlikely to become more complicated but the total amount of ventilation allowed for when fans are used has often been extremely sparing, and a full scale of air flow, up to 0·50 cubic feet per minute per lb liveweight, will ensure correct conditions are always maintained even in the warmest areas.

BETTER INSULATION

Though the present standard of insulation as used in the better piggeries is good, likewise we shall see an even higher one used as farmers appreciate the great practical benefits of the same, and the remarkably small increase in total cost that is entailed. For example, when roof insulation was first used a standard of 1″ thickness of glass or mineral wool was considered sufficient. From there we have increased to 2″ as being a normally acceptable standard, but any piggery would benefit greatly from as much as 4″ of glass or mineral wool correctly vapour-sealed. At Cambridge we erected a piggery 13 years ago with this standard of insulation, and the benefits have been immense; it is almost impossible to obtain condensation on the surfaces and the heat loss and heat gain are so reduced that temperature stability is easily maintained. In a building that cost approximately £1 per square foot the extra cost of using 4″ mineral wool instead of 1″ was only 3p so no one can reject its use on the grounds of excessive cost!

In our determination to provide warmth at the minimum cost, the cubic area of all piggeries will be reduced or special kennels will be provided as a practical alternative in the naturally-ventiated building. It is unlikely that we shall see any increase in the number of pigs kept in one pen—that is 10 to 20 per pen of fatteners. Already there is evidence to suggest that large groups in the strawless and slatted-floor units find it difficult to settle comfortably and unevenness of growth, fighting and cannibalism may result. The minimum area we shall accept will be in the order of 4–4½ square feet for porkers, 5 square feet for baconers, and 5½ square feet for heavy hogs. But there is no harm in allowing slightly more if possible; as wise as it is to pack pigs in, they must not be overcrowded and there has been a tendency to go to that extreme recently.

NEW SHAPES AND FORMS

New shapes and forms will also have some part to play. In economising on space and easing climate control, great possibilities exist for multi-storey buildings and I would like to see as much interest taken in them here as in other countries. Some of the advantages that are apparent are as follows. Provided that the correct design is used, a cheapening of constructional costs; traditional forms tend to make two-storey buildings more expensive, but with prefabrication in lighter materials the cost can undoubtedly be less. With multi-storey buildings, only one foundation is needed for two or three floors and insulation is only required in the base floor. Also, there is only one roof, which is the source of the greatest heat loss, so that by eliminating one or two roofs in two- or three-storey buildings, we have reduced costs and given considerable assistance in maintaining both uniform and stable conditions. Further, the amount of land needed is less, the services are all centralised, and with mechanical means to take the food, stock and litter upwards, the distribution can be downwards, thereafter, by gravity. Maintenance costs are also going to be less, particularly as the exposed surface area of the building is so much reduced. One possible snag is in arranging depopulation and disinfection—by concentrating the stock it is certain to be more difficult. Provided the building is not too large, each floor can be considered a separate house.

The round piggery also has attractive potential of a less ambitious sort. One of the best (designs) is in line with many of the trends we have already predicted. It consists of a 30′

diameter building with troughs or feeders round the passage and a central slatted dunging area. The pens divisions radiate from the centre outwards and revolve, offering a unique system of adjusting the pen size according to the age and/or number of pigs housed. Feeding is carried out by any system chosen. The dung falls into a sludge pit under the slats and is pumped out from here.

There are several advantages to this arrangement. A piggery of 30′ diameter takes 180 pigs, which is a satisfactory small number for a building from the point of view of disease control. Ventilation is cheap with inlets around the walls and one central extractor unit over the dunging area. Floor space per pig is minimal, but trough space, being at the outside of the building, is kept to a high figure. The dunging area is also reduced to a minimum and the construction of slats is on the most economical lines. Round the buildings offer scope in economy of walling, traditional buildings having up to twice the wall surface, roof and floor areas. In addition a two-storey round piggery adapted from a grain-storage bin and of American design has been marketed here. It is 48′ in diameter with a capacity of 47 sows and 450 bacon pigs and has automatic ventilation, feeding and dung clearing. Small multi-storey circular buildings may be an interesting development for the more adventurous, offering a saving in capital, maintenance and labour costs.

SPECIALISATION

Two opposing lines of thought exist on the question of whether a building should be specialised or not. Personally, I believe that pigs, or for that matter any species of livestock, require such specialisation in design that it is generally impossible to have a good pig-house and a building that can be cheaply adapted to other livestock requirements. Perhaps those who think in terms of unspecialised buildings know too little of the pig's requirements. Certainly the most successful producers now build for pigs alone and thus take advantage of every advance that is made in the most economical way possible. The mere operation of converting a building from pigs to other uses appears extremely wasteful. The cost of the shell may represent only half the cost of the building; the rest—in trough, pen divisions, gates, specialised floors, ventilation, dividing walls, etc—will have to be removed for its efficient use for other stock. Into this shell the divisions and fittings have to be placed for the new species. Anyone erecting a piggery should have more confidence in the industry and in his own ability than to think along these lines.

The problem remains, however, that buildings erected today will inevitably be out of date within a few years. Traditional methods of erection give a construction which lasts far too long, and thus outlives its useful life. The answer must surely be in the development of highly-specialised buildings giving us every economy that modern advances make possible, but with a limited life. A maintenance-free existence of 15 to 20 years is quite enough to plan for most piggeries. We can only hope that the pig industry will co-operate with the building interests to produce a range of designs that will answer these requirements, and by carrying out the job on factory lines, will reduce the cost to a minimum.

INFLUENCE OF CAGES ON EARLY WEANING

In the last twenty years there have been several attempts at improving productivity by early weaning piglets, from 4 days onwards, and usually into cages. These have not always been successful, but recent experience in the USA, Western Germany and Belgium has culminated in the development of a British system referred to earlier in the book (page 131) which promises a greater measure of success.

If the breeding pigs will produce the extra piglets to justify the enthusiastic response, then there will undoubtedly be a marked improvement in productivity together with a greater trend towards specialisation, which this system demands, and a further growth of larger units.

The striking growth of large and standardised units in the East Riding of Yorkshire shows a pattern of intensification that we may well see copied elsewhere. The Yorkshire system utilises prefabricated buildings throughout, and manages the sows individually by one of the systems described in Chapter 10. Weaning is usually into a verandah-type weaner house with weldmesh floor (page 132) at an age between 3 and 6 weeks depending on the husbandry system. From here the piglets go to the fattening house (often slatted with floor feeding) between 80 and 100 lb. Muck is usually disposed of as slurry and a minimum of straw is used. Early weaning in cages would fit in well with this system.

In the next decade one can anticipate improved breeding, feeding and management techniques that will demand a greater efficiency than ever in housing. I believe we have the knowledge available to provide this and indeed that the pattern can be found developing now at a fast yet sensible pace.

APPENDICES

Appendix 1

SUMMARY OF ENVIRONMENTAL REQUIREMENTS

TEMPERATURES:	Early weaned pigs	80–85°F
	Piglets in nest	70–80°F
	Farrowing sows	50°F minimum
		60°F ideal
	Sows in stalls	55°F minimum
	Fatteners: Weaners	70–75°F
	Stores	65–70°F
	Porkers	65–70°F
	Baconers	60–70°F
	Heavy hogs	50–60°F

When bedding is used temperatures at least 10°F below the above are acceptable.

HUMIDITY: Generally aim for a maximum of 70 to 80 per cent and a minimum of 40 per cent.

VENTILATION:

Farrowing House. Maximum for sow and litter of 250 cubic feet per minute or 20 air changes whichever is the least. Winter rate may be reduced to one-tenth of the maximum, which is 25 cfm for the sow and her new-born litter.

Fattening. Maximum rate for summer months should be $\frac{1}{2}$ cfm per lb bodyweight, that is, 120 cfm per heavy pig, 100 per baconer and 40 cfm per porker. Winter rates can be reduced to a fifth of this rate which will disperse all the moisture produced by the pigs in the coldest weather. In cooler areas of the country (North and North-east) maximum rates may be half those quoted.

LIGHTING:

Covered Yards. Roof lighting in covered yards which are uninsulated can be provided by using translucent sheeting at one-twentieth of the floor area. Use preferably on north-facing slopes only.

Farrowing and Fattening. Roof lights are best avoided as they cause unnecessary condensation and heat loss or gain even if double-glazed. Window lighting can be provided satisfactorily by installing windows at the rate of one-tenth of the wall area. Thus in a wall section 6' high and 10' long, a window would be placed 2' high and 3' long. This should be placed 1' below the eaves and where used as ventilation as well, be of the inward-opening hopper type. All windows must be double-glazed. Natural lighting is not essential and the construction of the building and its environment qualities are improved if it is omitted.

Artificial Lighting is most satisfactorily provided by using tungsten bulb points at 10' centres. Low intensities can be used continuously over dung passages and higher intensities at the rate of $\frac{3}{4}$ watt per sq ft of floor space where intermittent lighting is needed. To control the behaviour of the pigs a voltage regulator in the lighting circuit is useful.

201

Appendix 1 (*continued*)

ARTIFICIAL HEATING
Farrowing Pens

For the farrowing sow and litter—infra-red lamps over nest: 500 watts.
For general space-heating six sows and litter to each 3 kW space-heater.

Fatteners

To reduce 'stress' of change apply $\frac{1}{12}$ kW per weaner installed for a few weeks in cold weather. Use a thermostat for control (pigs produce heat at the rate of approximately 5 BTU per lb bodyweight, or, at bacon weight, pig produces $\frac{1}{5}$ kW).

Appendix 2

SUMMARY OF MEASUREMENTS

TROUGH SPACE (*Minimum*)		
Weaner	6 inches	
Porker	10 inches	
Baconer	12 inches	
Heavy hog	14 inches	
Sow	18 inches	

AREAS

Farrowing and Rearing	Total area	70–100 sq ft
	Divided in some cases between:	
	Sleeping and feeding area	40–60 sq ft
	Dunging area	30–40 sq ft
	Creep area	15 sq ft minimum
	Creep area within 'sleeping' area: frontage of creep	6 ft
Farrowing Crates	Total area up to	70 sq ft
	Stall	40 sq ft minimum
	Optional exercising area	30–40 sq ft
Early Weaner Pens (2–8 weeks)	Total area up to	$2\frac{1}{2}$ sq ft per pig
	Sleeping and feeding	$1\frac{3}{4}$ sq ft per pig
	Dunging area	$\frac{3}{4}$ sq ft per gig
Intensive Weaner Pens (8-12 weeks)	Total area	$3\frac{1}{2}$ sq ft per pig
	Sleeping and feeding	$2\frac{1}{2}$ sq ft per pig
	Dunging area	1 sq ft per pig
Straw-yard for Weaners (6–14 weeks)	Total area up to	10 sq ft per pig
Intensive Growing Pens (Stores or Porkers —12–18 weeks)	Total area	5 sq ft per pig
	Sleeping and feeding	$3\frac{1}{2}$ sq ft per pig
	Dunging	$1\frac{1}{2}$ sq ft per pig
Baconers	Total area	7 sq ft per pig
	Sleeping and feeding	5 sq ft per pig
	Dunging	2 sq ft per pig
'*Heavies*'	Total area	$7\frac{1}{2}$ sq ft per pig
	Sleeping and feeding	$5\frac{1}{2}$ sq ft per pig
	Dunging	2 sq ft per pig

Appendix 3

SUMMARY OF COSTS

Estimates of costs are necessarily approximate as so much depends on the area in which the building is situated and the materials and the site conditions. The costs here are based on actual cases in several areas for building erection. As labour represents about two-thirds of the cost and materials one-third considerable saving may be obviously made if one's own labour is used.

Farrowing	*Cost per pen*
Totally-enclosed farrowing pen	£200 to £240
Single farrowing pen with outside access	£100 to £120
Farrowing huts	£70 to £100

Fattening
Conventional Design

Totally-enclosed weaner pen	up to £14 per pig
Totally-enclosed grower or porker pen	up to £17 per pig
Totally-enclosed baconer pen	up to £22 per pig
Totally-enclosed heavy hog pen	up to £22 per pig
Pens with outside covered yards	£22 per pig
Pens with outside uncovered yards	up to £17 per pig
Kennel pen and covered yards	up to £16 per pig
Simple 'loose-box' type accommodation with fatteners at 20–30 per unit	up to £10 per pig

By omitting troughs and feeding on the floor from overhead catwalks, costs may be reduced by one-third to a half. Costs also depend on numbers of pigs per pen and the highest cost is with small numbers, i.e. ten pigs per pen.

Appendix 4
METRIC CONVERSION FACTORS

Definition	To Convert	into	multiply by
Temperature rise	deg F	deg C	0·55
	deg C	deg F	1·8
Length	in	mm	25·400
	mm.	in	0·0394
	ft	metres (m)	0·3048
	m	ft	3·2808
Area	ft^2	m^2	0·0929
	m^2	ft^2	10·7639
Volume	ft^3	m^3	0·0283
	m^3	ft^3	35·3148
Velocity	ft/min	m/sec	0·0051
	m/sec	ft/min	196·8504
Mass	lb	Kg	0·4536
	Kg	lb	2·2046
Rate of thermal	Watts (W)	Btu/h	3·4121
transmission	Btu/h	W	0·2931
	W	kcal/h	0·8598
	kcal/h	W	1·1630
	Btu/h	kcal/h	0·2520
	kcal/h	Btu/h	3·9680
	Btu/h/deg F	kcal/h/deg C	0·4536
	kcal/h/deg C	Btu/h/deg F	2·2044

Building heat loss data

Thermal conductivity	Btu in/sq ft h deg F (K value)	kcal/m h deg C	0·1240
	kcal/m h deg C	Btu in/sq ft h deg F	8·0636
Thermal conductance and thermal trans- mittance	Btu/sq ft h deg F (U-value)	kcal/sq m h deg C	4·8824
	kcal/sq m h deg C	Btu/sq ft h deg F	0·2048
Thermal resistivity	sq ft h deg f/Btu in (1/K)	sq m h deg C/kcal m	8·0636
	sq m h deg C/kcal m	sq ft h deg F/Btu in	0·1240
Thermal resistance	sq ft h deg F/Btu	sq m h deg C/kcal	0·2048
	sq m h deg C/kcal	sq ft h deg F/Btu	4·8824
	deg F/Btu/h	deg C/kcal/h	2·2044
	deg C/kcal/h	deg F/Btu/h	0·4536

Ventilation heat loss data

Volume rate of flow	ft^3/min	m^3/h	1·699
	m^3/h	ft^3/min	0·5886
Approx rate of thermal extraction by air	ft^3/min	Watts/deg F	0·317
	m^3/h	W/deg F	0·1866
	m^3/h	W/deg. C	0·3358
	ft^3/min	Btu/h/deg F	1·08
	m^3/h	Btu/h/deg F	0·637
	m^3/h	kcal/h/deg C	0·2886

Above figures are calculated on basis of weight of dry air being 0·075 lb/ft^3 (1·2 Kg/m^3) and specific heat of dry air of 0·24 Btu/lb (0·133 kcal/Kg).

Atmospheric water vapour data

Moisture respiration rate	grains/hr (gr/h)	gms/hr (g/h)	0·0648
	g/h	gr/h	15·432
Water vapour con- centration in at- mosphere by weight	gr/lb of air	g/Kg of air	0·143
	g/Kg of air	gr/lb of air	7·0
Water vapour concen- tration in atmos- phere by volume	gr/ft^3 of air	g/m^3 of air	2·285
	g/m^3 of air	gr/ft^3 of air	0·437

Appendix 5

TEMPERATURE CONVERSION TABLE

°F	°C	°F	°C
20	−6·7	66	18·9
21	−6·1	67	19·4
22	−5·6	68	20·0
23	−5·0	69	20·6
24	−4·4	70	21·1
25	−3·9	71	21·7
26	−3·3	72	22·2
27	−2·8	73	22·8
28	−2·2	74	23·3
29	−1·7	75	23·9
30	−1·1	76	24·4
31	−0·6	77	25·0
32	0	78	25·6
33	0·6	79	26·1
34	1·1	80	26·7
35	1·7	81	27·2
36	2·2	82	27·8
37	2·8	83	28·3
38	3·3	84	28·9
39	3·9	85	29·4
40	4·4	86	30·0
41	5·0	87	30·6
42	5·6	88	31·1
43	6·1	89	31·7
44	6·7	90	32·2
45	7·2	91	32·8
46	7·8	92	33·3
47	8·3	93	33·9
48	8·9	94	34·4
49	9·4	95	35·0
50	10·0	96	35·6
51	10·6	97	36·1
52	11·1	98	36·7
53	11·7	99	37·2
54	12·2	100	37·8
55	12·8	101	38·3
56	13·3	102	38·9
57	13·9	103	39·4
58	14·4	104	40·0
59	15·0	105	40·6
60	15·6	106	41·1
61	16·1	107	41·7
62	16·7	108	42·2
63	17·2	109	42·8
64	17·8	110	43·3
65	18·3		

Appendix 6

CODE OF RECOMMENDATIONS FOR THE WELFARE OF PIGS
INTRODUCTION

1. The welfare of pigs can be safeguarded under a variety of management systems. The system, and the number and density of pigs kept at any one time, should depend on the suitability of the accommodation and the skills of the stockman. Badly-managed and unhealthy pigs will not do well, and it is essential that the stockman should watch for signs of distress or disease.
2. The good stockman will know the signs which indicate good health in pigs. He should be able to recognise impending trouble in its earliest stages and may often be able to identify the cause and put matters right immediately. If the cause is not obvious, or if the stockman's immediate action is not effective, veterinary advice should be obtained as soon as possible.
3. The signs of ill-health in pigs include separation from the group, lameness, poor appetite, vomiting, constipation, diarrhoea, discolouration of the skin, shivering, sneezing, persistent coughing or panting, and swelling of the throat, navel, udder or joints.

GENERAL RECOMMENDATIONS: ALL PIGS
HOUSING

4. Advice on welfare aspects should be sought when new buildings are to be constructed or existing buildings modified.
5. Internal surfaces of housing and pens should be of materials which can be effectively cleansed and disinfected, or which are disposable.
6. Internal surfaces and fittings of buildings and pens accessible to pigs should not have sharp edges or projections likely to cause injury.
7. Pen floors should be effectively drained. All floors, particularly slatted or metal mesh ones, should be designed, constructed and maintained so as to avoid injury or distress to the pigs. Advice should be sought if injury or distress occurs.
8. Paints and wood preservatives which may be toxic to pigs should not be used on surfaces accessible to them. Particular care is necessary to guard against the risk of poisoning from old paint-work in any part of the building or when second-hand building materials are used.
9. When planning new buildings, consideration should be given to the provision of an escape route for stock in an emergency; and materials used in construction should have sufficient fire resistance to enable any emergency procedure to be followed.

VENTILATION AND TEMPERATURE

10. Excessive heat loss should be prevented either by the structural insulation of the external walls, roof and floor of the lying area or by the provision of adequate bedding. Sufficient ventilation is essential.
11. When removing slurry from under slats, special care is essential to avoid fouling the air with dangerous gases which may be fatal to man and animals.

LIGHTING

12. Throughout the hours of daylight the level of indoor lighting, natural or artificial, should be such that all housed pigs can be seen clearly. In addition, adequate lighting should be available for satisfactory inspection at any time.

MECHANICAL EQUIPMENT AND SERVICES

13. Essential mechanical equipment, such as feeding hoppers, drinkers, ventilator fans, heating and lighting units, fire extinguishers and alarm systems, should be inspected regularly and kept in good working order. There should be an alarm system to warn of failure of any essential automated equipment. Alternative ways of feeding and of maintaining a satisfactory environment should be available for use in the event of a breakdown. All electrical installations at mains voltage should be inaccessible to pigs and properly earthed.

SPACE ALLOWANCES

14. When pigs are fed by any system which not allow continuous and unrestricted access to food all pigs in the group should be able to feed at the same time.

FOOD AND WATER

15. Whatever feeding system is adopted, all pigs should receive a daily diet which is nutritionally adequate to maintain health.
16. Fresh clean water or other wholesone liquid should be available to the pigs daily.
17. When sows are being dried off they should receive food and water at intervals of not more than 24 hours.

MANAGEMENT

18. Pigs should be closely inspected at least daily, preferably when feeding, for signs of injury, illness or distress.
19. Extremes of air temperature or of humidity, particularly those liable to cause heat stress, should not be deliberately maintained.
20. Sick or injured pigs should be treated without delay. Accommodation should be available to enable them to be isolated.
21. Where it is necessary to mark pigs for permanent identification, the ear may be tattooed, tagged, notched or punched. Slap marking is an acceptable method where identification is required for a temporary period only. These operations should be carried out by competent operators, exercising care to avoid unnecessary pain or unnecessary distress to the pigs.
22. Docking should not be carried out unless prescribed by a veterinary surgeon.

ADDITIONAL RECOMMENDATIONS: PIGS KEPT INDOORS

FARROWING PIGS AND SUCKLING PIGLETS

23. Farrowing quarters should have a farrowing crate, farrowing rails or other device for the protection of the piglets.
24. Sows should be placed in clean and comfortable farrowing quarters well before the litter is due.
25. A temperature suitable for the piglets should be maintained in the nest or creep, either by insulation or by artificial heating. This temperature should be significantly higher than that provided for the sow. Heating devices should be protected from interference by the sow or piglets.

GROWING PIGS

26. The total floor space should be adequate for sleeping and feeding and of such size that soiling of the lying area may be avoided. As a guide, fatteners and maiden gilts above 57 kg (125 lb) liveweight should be allowed a minimum accessible area (including dunging area) of 1 sq m for each 122 kg (1 sq ft for each 25 lb) liveweight. Proportionately more space should be allowed for pigs of lower weights, particularly when kept in straw yards.

BREEDING SOWS AND GILTS

27. Aggressiveness in dry sows presents a severe problem of husbandry. Where the sows or gilts are kept in groups, much depends on the temperament of individual animals; but the stockman should ensure that persistent bullying leading to severe injury does not take place.
28. Where sows or gilts are housed individually they should be able to feed and lie down normally. The provision of bedding may be advantageous. Partitions between pens should be designed so that fighting cannot occur, and should allow the animals to see each other. If tethers are used, they should not cause injury or distress.

BOARS

29. As a guide, individual accommodation for an adult boar should have a floor area of not less than 7 sq m (75 sq ft) if used for living purposes only. If used for both living and service purposes the floor area should be not less than 9·3 sq m (100 sq ft) with the shortest side not less than 2·1 m (7 ft). In both cases the pen divisions should be not less than 1·4 m (4 ft 6 in) high.
30. In a single-purpose pen, bedding should be provided in the lying area. In a dual-purpose pen, an adequate part of the floor area should be bedded, and the whole floor area should be kept dry, or sufficient bedding provided to give an adequate grip during service. The use of a service crate may be advantageous.
31. The trimming of boars' tusks is advisable where injury to man or animals is likely to occur. It may be carried out by a competent person following veterinary guidance.

ADDITIONAL RECOMMENDATIONS: PIGS KEPT OUTDOORS

32. Huts used for farrowing and rearing should have a warm and draught-free bed for the sow and litter.
33. When sows are tethered. the design and length of tether should be such as to prevent sows becoming entangled either with the huts or with each other.
34. Adequate shelter in winter and shade in summer should be available to all pigs.

INDEX